Innovative
Teaching

**Best Practices from Business and
Beyond for Mathematics Teachers**

Problem Solving in Mathematics and Beyond

Print ISSN: 2591-7234
Online ISSN: 2591-7242

Series Editor: Dr. Alfred S. Posamentier
Distinguished Lecturer
New York City College of Technology - City University of New York

There are countless applications that would be considered problem solving in mathematics and beyond. One could even argue that most of mathematics in one way or another involves solving problems. However, this series is intended to be of interest to the general audience with the sole purpose of demonstrating the power and beauty of mathematics through clever problem-solving experiences.

Each of the books will be aimed at the general audience, which implies that the writing level will be such that it will not engulfed in technical language — rather the language will be simple everyday language so that the focus can remain on the content and not be distracted by unnecessarily sophiscated language. Again, the primary purpose of this series is to approach the topic of mathematics problem-solving in a most appealing and attractive way in order to win more of the general public to appreciate his most important subject rather than to fear it. At the same time we expect that professionals in the scientific community will also find these books attractive, as they will provide many entertaining surprises for the unsuspecting reader.

Published

Vol. 24 *Innovative Teaching: Best Practices from Business and Beyond for Mathematics Teachers*
by Denise H Sutton and Alfred S Posamentier

Vol. 23 *Learning Trigonometry by Problem Solving*
by Alexander Rozenblyum and Leonid Rozenblyum

Vol. 22 *Mathematical Labyrinths. Pathfinding*
by Boris Pritsker

Vol. 21 *Adventures in Recreational Mathematics: Selected Writings on Recreational Mathematics and its History*
(In 2 Volumes)
by David Singmaster

For the complete list of volumes in this series, please visit www.worldscientific.com/series/psmb

Problem Solving in
Mathematics and Beyond

Volume **24**

Innovative Teaching

Best Practices from Business and Beyond for Mathematics Teachers

Denise H. Sutton
Alfred S. Posamentier

City University of New York, USA

World Scientific

NEW JERSEY · LONDON · SINGAPORE · BEIJING · SHANGHAI · HONG KONG · TAIPEI · CHENNAI · TOKYO

Published by

World Scientific Publishing Co. Pte. Ltd.

5 Toh Tuck Link, Singapore 596224

USA office: 27 Warren Street, Suite 401-402, Hackensack, NJ 07601

UK office: 57 Shelton Street, Covent Garden, London WC2H 9HE

Library of Congress Control Number: 2021058786

British Library Cataloguing-in-Publication Data
A catalogue record for this book is available from the British Library.

Problem Solving in Mathematics and Beyond — Vol. 24
INNOVATIVE TEACHING
Best Practices from Business and Beyond for Mathematics Teachers

ISBN 978-981-123-166-7 (hardcover)
ISBN 978-981-123-167-4 (ebook for institutions)
ISBN 978-981-123-168-1 (ebook for individuals)

For any available supplementary material, please visit
https://www.worldscienti ic.com/worldscibooks/10.1142/12144#t=suppl

Desk Editors: Vishnu Mohan/Tan Rok Ting

Typeset by Stallion Press
Email: enquiries@stallionpress.com

Printed in Singapore

About the Authors

 **Denise H. Sutton, PhD**, is an Assistant Professor at New York City College of Technology—The City University of New York (CUNY), where she teaches business history, marketing, and fashion. She is an Adjunct Associate Professor at the Fashion Institute of Technology—SUNY, in the Cosmetics and Fragrance Marketing and Management program, School of Graduate Studies, where she teaches a course on the history of innovation in the beauty industry. Dr. Sutton also developed and taught courses in advertising and gender at the New School University, New York City, and gender studies courses at Clark University, Worcester, MA. She is the author of *Globalizing Ideal Beauty: Women, Advertising, and the Power of Marketing* (Palgrave Macmillan 2009, 2012).

Dr. Sutton is a 2020 Fulbright Scholarship recipient and will present a series of lectures on the topic of innovation in teaching, including adapting best practices from business, at the University College of Teacher Education–Lower Austria. An expert on innovation and advertising, she has lectured widely on the subject at universities and at corporations such as Unilever and Firmenich; her research has been the subject of the annual research seminar series at the Hagley Museum's Center for Business, Technology, and Society. At the New

York City College of Technology, she is the Director of the Entrepreneurship & Innovation Program, and, as part of the program, developed the Smart Tank student pitch competition, based on what actually happens in the world of marketing and advertising. As an advisory board member of CUNY Startups, Dr. Sutton supports the organization's mission to improve the economic mobility of CUNY students through entrepreneurship while stimulating business growth and economic development in New York City.

Prior to her career in academia, Dr. Sutton held positions as director of communications at CUNY and at the Harlem Children's Zone (HCZ), where she worked for education pioneer Geoffrey Canada on education and community-building media campaigns and managed all aspects of the agency's image—locally and nationally—as defined by its mission. While at HCZ, she also contributed to the development of the Harvard Business School case study, "The Harlem Children's Zone: Driving Performance with Measurement and Evaluation."

Dr. Sutton earned her doctorate in Gender and Women's Studies at Clark University in Worcester, Massachusetts, where she focused on interdisciplinary research in communications, advertising, and history. She has a Master's Degree in 18th-Century British Literature from East Carolina University and a baccalaureate degree in English from the University of North Carolina, Wilmington.

 **Alfred S. Posamentier, PhD**, is currently a Distinguished Lecturer at New York City College of Technology of the City University of New York. Prior to that he was Executive Director for Internationalization and Funded Programs at Long Island University, New York. This was preceded by five years as Dean of the School of Education and Professor of Mathematics Education at Mercy College, New York. Before that he was for 40 years at The City College of the City University of New York, at which he is now Professor Emeritus of Mathematics Education and Dean Emeritus of the School of Education. He is the author and co-author of more than 75 mathematics books for

teachers, secondary and elementary school students, as well as the general readership. Dr. Posamentier is also a frequent commentator in newspapers and journals on topics related to education.

After completing his B.A. degree in mathematics at Hunter College of the City University of New York, he took a position as a teacher of mathematics at Theodore Roosevelt High School (Bronx, New York), where he focused his attention on improving the students' problem-solving skills and at the same time enriching their instruction far beyond what the traditional textbooks offered. During his six-year tenure there, he also developed the school's first mathematics teams (both at the junior and senior level). He is still involved in working with mathematics teachers and supervisors, nationally and internationally, to help them maximize their effectiveness.

Immediately upon joining the faculty of the City College of New York in 1970 (after having received his Master's degree there in 1966), he began to develop in-service courses for secondary school mathematics teachers, including such special areas as recreational mathematics and problem solving in mathematics. As Dean of the City College School of Education for 10 years, his scope of interest in educational issues covered the full gamut of educational issues. During his tenure as dean, he took the School from the bottom of the New York State rankings to the top with a perfect NCATE accreditation assessment in 2009. Dr. Posamentier repeated this successful transition at Mercy College, which was then the only college to have received both NCATE and TEAC accreditation simultaneously.

In 1973, Dr. Posamentier received his Ph.D. from Fordham University (New York) in mathematics education and has since extended his reputation in mathematics education to Europe. He has been visiting professor at several European universities in Austria, England, Germany, The Czech Republic, Turkey, and Poland. In 1990, he was Fulbright Professor at the University of Vienna.

In 1989, he was awarded an Honorary Fellow position at the South Bank University (London, England). In recognition of his outstanding teaching, the City College Alumni Association named him Educator of the Year in 1994, and in 2009. New York City had the day, May 1, 1994, named in his honor by the President of the New York viii *Innovative Teaching*

City Council. In 1994, he was also awarded the *Das Grosse Ehrenzeichen für Verdienste um die Republik* Österreich (Grand Medal of Honor from the Republic of Austria), and in 1999, upon approval of Parliament, the President of the Republic of Austria awarded him the title of University Professor of Austria. In 2003, he was awarded the title of *Ehrenbürgerschaft* (Honorary Fellow) of the Vienna University of Technology, and in 2004 he was awarded the *Österreichisches Ehrenkreuz für Wissenschaft & Kunst 1. Klasse* (Austrian Cross of Honor for Arts and Science, First Class) from the President of the Republic of Austria. In 2005, he was inducted into the Hunter College Alumni Hall of Fame, and in 2006 he was awarded the prestigious Townsend Harris Medal by the City College Alumni Association. He was inducted into the New York State Mathematics Educator's Hall of Fame in 2009, and in 2010 he was awarded the coveted Christian-Peter-Beuth Prize from the Technische Fachhochschule—Berlin. In 2017, Dr. Posamentier was awarded *Summa Cum Laude nemmine discrepante,* by the Fundacion Sebastian, A.C., Mexico City, Mexico.

He has taken on numerous important leadership positions in mathematics education locally. He was a member of the New York State Education Commissioner's Blue Ribbon Panel on the Math-A Regents Exams, and the Commissioner's Mathematics Standards Committee, which redefined the Mathematics Standards for New York State, and he also served on the New York City schools' Chancellor's Math Advisory Panel.

Dr. Posamentier is still a leading commentator on educational issues and continues his long time passion of seeking ways to make mathematics interesting to both teachers, students, and the general public—as can be seen from some of his more recent books: *Teaching Secondary School Mathematics: Techniques and Enrichment* (World Scientific Publishing, 2020), *The Joy of Geometry* (Prometheus Books, 2020), *Mathematics Entertainment for the Millions* (World Scientific Publishing, 2020), *Math Makers: the Lives and Works of 50 Famous Mathematicians* (Prometheus Books, 2020), *Understanding Mathematics Through Problem Solving* (World Scientific Publishing, 2020), *The Psychology of Problem Solving: The Background to Successful*

Mathematics Thinking (World Scientific Publishing, 2020), *Solving Problems in Our Spatial World* (World Scientific Publishing, 2019), *Tools to Help Your Children Learn Math: Strategies, Curiosities, and Stories to Make Math Fun for Parents and Children* (World Scientific Publishing, 2019), (Prometheus, 2018), *The Joy of Mathematics* (Prometheus Books, 2017), *Strategy Games to Enhance Problem-Solving Ability in Mathematics* (World Scientific Publishing, 2017), *The Circle: A Mathematical Exploration Beyond the Line* (Prometheus Books, 2016), *Effective Techniques to Motivate Mathematics Instruction*, 2nd Ed. (Routledge, 2016), *Problem-Solving Strategies in Mathematics* (World Scientific Publishing, 2015), *Numbers: There Tales, Types, and Treasures* (Prometheus Books, 2015), *Mathematical Curiosities* (Prometheus Books, 2014), *Magnificent Mistakes in Mathematics* (Prometheus Books, 2013), 100 *Commonly Asked Questions in Math Class: Answers that Promote Mathematical Understanding, Grades 6–12* (Corwin, 2013), *What Successful Math Teachers do—Grades 6–12* (Corwin, 2013), *The Secrets of Triangles: A Mathematical Journey* (Prometheus Books, 2012), *The Glorious Golden Ratio* (Prometheus Books, 2012), *The Art of Motivating Students for Mathematics Instruction* (McGraw-Hill, 2011), *The Pythagorean Theorem: Its Power and Glory* (Prometheus, 2010), *Teaching Secondary Mathematics: Techniques and Enrichment Units*, 9th Ed. (Pearson, 2015), *Mathematical Amazements and Surprises: Fascinating Figures and Noteworthy Numbers* (Prometheus, 2009), *Problem Solving in Mathematics: Grades 3–6: Powerful Strategies to Deepen Understanding* (Corwin, 2009), *Problem-solving Strategies for Efficient and Elegant Solutions, Grades 6–12* (Corwin, 2008), *The Fabulous Fibonacci Numbers* (Prometheus Books, 2007), *Progress in Mathematics K-9* textbook series (Sadlier-Oxford, 2006–2009), *What successful Math Teachers Do: Grades K-5* (Corwin, 2007), *Exemplary Practices for Secondary Math Teachers* (ASCD, 2007), *101+ Great Ideas to Introduce Key Concepts in Mathematics* (Corwin, 2006), π, *A Biography of the World's Most Mysterious Number* (Prometheus Books, 2004), *Math Wonders: To Inspire Teachers and Students* (ASCD, 2003), and *Math Charmers: Tantalizing Tidbits for the Mind* (Prometheus Books, 2003).

Contents

About the Authors   v

Introduction   xiii

Chapter 1   Know Yourself   1

Chapter 2   Know Your Audience   21

Chapter 3   Know How to Reach Your Students   43

Chapter 4   Know What Motivates Your Students   59

Chapter 5   Know How to Engage Your Students   103

Chapter 6   Empower Students to Become Leaders   119

Chapter 7   Keeping Up with Teaching Technologies   137

Chapter 8   Navigating School Culture   149

Chapter 9   Developing Leadership Skills   167

Chapter 10 Understanding the Changing Landscape
            of Gender and Race   179

Epilogue   195

Index   197

Introduction

The importance of innovation in the business world and beyond is a given. The same attention must be paid to innovation when it comes to teaching. Our unique approach in *Innovative Teaching: Best Practices from Business & Beyond for Mathematics Teachers* is to provide educators with exciting ways to introduce material and methods to motivate and engage students by showing how some of the techniques commonly used in the business world—and in fields such as journalism, medicine, technology, coaching, marketing, and sales, among others—are applicable to the world of education. To innovate is not to invent. One way to innovate in the classroom is to simply use an approach that recombines existing best practices from disparate fields and adapt those resources to a classroom setting. *Innovative Teaching* also offers educators practical advice with regard to the changing culture of education, including: keeping up with technology, navigating work culture, interacting with colleagues, developing leadership skills, and staying up-to-date with discussions and new terminology related to sexual orientation and gender identity.

Innovative Teaching demonstrates how the classroom environment is similar to the marketplace in certain respects; however, we recognize that *education is not a business* and that *students are not consumers*. Rather, our point suggests that teachers can tap into a rich world of best practices from fields beyond education and carefully choose the techniques and methods that best fit the needs of their own particular students. Educators, like businesses, for example, must capture and hold the attention of their audience while competing with a constant stream of "noise." With the introduction of the Internet and the wide use of social media, promoters understand that they must not only engage their audience, but they must incorporate audience feedback into the promotional work and product or service they offer. *Innovative Teaching* shows educators how to take some best practices from marketing and recombine these resources for appropriate use in the classroom.

Just as those in the business world take advantage of their personal strengths, educators should understand how their personality determines the approaches they may or may not take in the classroom. Because knowing oneself when engaged in a teaching career is fundamental to success, *Innovative Teaching* begins by helping teachers understand how their personality traits may have a positive and/or negative effect on their interactions with students. Setting out with deliberate intentions—that result from self-knowledge—will put the progressive teacher light years ahead of their colleagues. A chapter on "Knowing Your Audience" explains the importance of understanding a target audience (students) quickly so that teachers can develop and use material that is relevant and will resonate with their students. Highlighting some of the best practices from marketing will help teachers identify techniques that they can incorporate into the classroom early in the year. Of course, much like public relations professionals and promoters, teachers need to reach students through compelling content.

From motivating to engaging and then empowering students, the chapters are organized in a linear fashion so that the reader becomes aware of how each topic builds upon the previous chapter. The second half of the book shifts slightly to consider the changing culture of education, while still providing challenging and exciting mathematics examples and best practices from various other fields. When a teacher is confronted with making decisions about technology in the classroom, for instance, what are the criteria for the effective use of technology? Which technology would be most appropriate to meet specific goals? How do we understand the historic role of technology in education, and what might we learn from a historical perspective? A chapter on school culture identifies the elements of a strong school culture and examines what contributes to a weak culture. Clear examples are provided along with suggestions for how a teacher can effectively function within each type of school culture. The chapter on leadership suggests that teachers should start to look for leadership opportunities early on in their career and start to develop leadership skills that align with their particular personality traits. By describing the most common leadership types, this chapter also provides insight into the personality traits of those who are in supervisory positions, enabling teachers to "manage up," a term that refers to the deft management of your supervisor. Last, but not least, the final chapter focuses on the topic of sexual orientation, gender identity, and the new terminology that every teacher should know. Many teachers do struggle with keeping a pace with the seemingly fast-changing nature of these important topics. However, it is worth the effort to fully understand the dynamics of gender identity and sexual orientation for the benefit of students and for teachers' own ability to create a safe classroom environment where learning can take place.

Innovative Teaching will help educators creatively explore options not offered through traditional education theories and/or curricula. Yet this approach does not suggest an overhaul of basic education curricula—it is adaptable for use in curriculum taught throughout the world, including current practices offered by the various state standards. Once educators understand innovation in the classroom, they can explore other fields and borrow best practices especially suited for their lessons and student needs.[1] The only limitation is the teacher's own imagination.

---

[1] One such book to prepare mathematics teachers at the secondary school level is *Teaching Secondary School Mathematics: Techniques and Enrichment Units*, by A. S. Posamentier and B. Smith, Hackensack, NJ: World Scientific Publishing, 2021, which provides all the necessary tools needed in the mathematics classroom as well as providing 125 enrichment units to bring teaching to a higher level.

Chapter 1

Know Yourself

Imagine the first day of school, standing in front of the classroom full of new students. You have prepared a meticulous lesson plan, and you have even found a delightful "ice-breaker" to help the students get to know each other and for you to get to know them. How much of the material presented in that first half an hour or 45 minutes do you think the students are able to retain? Most likely, very little. If the students are not absorbing the lesson or remembering classmates' names, what are they thinking about during that time? The simple answer is—*they are thinking about you*. Typically, students are trying to figure out what kind of a teacher you will be: Will you be strict? Will you be a push-over? Can they trust you? Are they going to have fun in the class? So, when you walk up to the front of the class (if this is the approach) on the first day, it is to your advantage to already know who you are as a teacher, to recognize where your strengths and weaknesses lie. You should also be aware that your students will be distracted by the completely understandable human attempt at trying to figure out who you are as a teacher and how you will handle your classroom. Your demeanor should communicate the message to these new students that they can be confident in the knowledge that your classroom will be a safe place where they can learn and thrive.

Just as those who work in other fields take advantage of personal traits to excel in their careers, as an educator you should understand how your personality can determine the approaches you may or may not take in the classroom. Your personality can determine teaching strategy and preferred methodology, which is fine, as long as you are aware of what is driving your decisions in the classroom. To be aware of your own particular personality strengths and weaknesses can be an advantage in teaching, and this awareness has the potential to enable you to improve or to adjust your performance as an educator. Awareness of self is key. You have the tremendous responsibility to guide students to learn and to develop as fully capable and expressive humans both emotionally and academically. In this sense, the focus should be on the students. However, as you start your teaching career, let us make sure you know who you are and what motivates you so that you can excel in your chosen profession.

Where do you begin your attempt at understanding who you are as teacher? First, think about the style of teaching to which you are "naturally" drawn. For instance, do you prefer a lecture-based approach for presenting material to your students? Or are you inclined to create small groups and encourage collaboration and discussion among students while you check in with each group on their progress? If you consider yourself an extrovert, are you more comfortable dealing with students who mirror your own tendencies or do you tend to favor students who exhibit the traits of an introvert? Do you feel comfortable with ambiguity or are you more interested in clear-cut, indisputable facts? Second, once you understand how your personality informs your teaching, you can practice making the necessary adjustments according to the needs of your students. If you are a high-energy lecturer who thrives on performance, for instance, you might want to learn how to develop the skills necessary to engage, motivate, and put the spotlight on students when appropriate.

Finally, think about how the perceptions of others have shaped you throughout your life. If you see yourself as an introvert, for example, that thinking may be influenced by how your family members described you as a child. Does that perception of you as a shy child still fit with who you are today? One way our personality traits are constructed is through the eyes of our family and friends as well as complete strangers. If throughout the course of a day you receive three compliments on a jacket you are wearing, that jacket might well become your favorite. The positive reinforcement of a compliment from co-workers or even a stranger you pass on the street will likely influence your perception of that jacket. While this example is not about a specific personality trait, it illustrates how feedback can impact our perception of an object (the jacket) that, through the language of fashion, communicates something about the wearer's personality. Society is that mirror through which we all see ourselves, and we are subjected to both positive and negative messages through our interaction with this mirror. This dynamic is what the social psychologist Charles Horton Cooley described in his theory of the "looking-glass self." Cooley suggests that we develop our sense of self through the perception and judgment of others.[1] Of course some people are more deeply affected than others by the perception and judgment of society, but all of us are affected to a degree. As you think more deeply about what kind of teacher you are (or will be), examine the assumptions you may have about who you are now as an adult. And remember, just as you have been influenced by the opinions and judgments of others, as a teacher, you will have that same influence on the students who observe your behavior and actions almost every day over the course of a year. Essentially, receiving compliments is in and of itself a very motivating factor, which should have positive results on a student's performance.

You do not have to be controlled by your inclinations or personality traits as they relate to teaching. In this chapter, we will explore some theories from the field of psychology to help gain an understanding of various educator personality types and what those types might look like in the classroom. We will also consider how the fields of theatre and journalism might inform the practice of teaching, especially those teaching approaches that may be outside of your comfortable personality type.

Suggested techniques for modifying and alternating teaching styles and approaches will be explored with a focus on what is commonly referred to as the Five-Factor Model. Studies on teacher

---

[1] C. H. Cooley, *Human Nature and Social Order*. New York, NY: Charles Scribner's Sons, 1922.

personality in educational psychology and related fields are rather diverse; however, the Five-Factor Model is one of the leading approaches to understanding personality in the field of psychology. It is a descriptive model that details personality traits along the following dimensions of (1) neuroticism (tendency toward anxiety and depression), (2) extraversion (social, assertive, opposite of introversion), (3) openness to experience (intellectual curiosity, creative imagination), (4) agreeableness (compassion, respectfulness, trust in others), and (5) conscientiousness (organized, productive, responsible).[2] Your personality type can lead to specific preferences when deciding, for instance, which teaching methodology to follow, or what style of teaching to adopt. It can be helpful to be aware of your own personality type, which can help you avoid suboptimal teaching practices and utilize best practices. The Five-Factor Model is one theory that may provide insight into personality traits of effective teachers.

Another measure of personality that you may be familiar with is the Myers–Briggs Type Inventory (MBTI),[3] which is a questionnaire based on Carl Jung's theory of psychological types. This theory suggests that there are four psychological functions through which the world is experienced: sensation, intuition, feeling, and thinking (expressed in either extrovert or introvert form). One of these functions is dominant and determines how people see the world and make decisions. Jung's theory is turned into practical use by using the MBTI to classify people (e.g. teachers) into types based on self-reported preferences.

In one revealing study, researchers used the MBTI to find out what personality traits make a good teacher. Their study of teacher personality found that a select group of teachers, rated the "best" teachers through other methods, who took the MBTI in the United States had a predominance of the "extraversion, intuition, feeling, and

---

[2] R.R. McCrae & P.T. Costa, A Five-Factor Theory of Personality. In L.A. Pervin & O.P. John (Eds.), *Handbook of Personality: Theory and Research*. New York, NY: Guilford Press, 1990, pp. 139–153.

[3] Take this test online, free of charge, at: https://www.16personalities.com.

perceiving" (ENFP) type. These are the particular types that represent the traits present in highly rated teachers. For "typical" or average teachers, a predominance of "extraversion, sensing, feeling, and judging" (ESFJ) types emerged. This study found that successful teachers were likely to be more oriented towards their environment, be inclined to make friends, think aloud and be rather active (extroversion). These teachers are oriented to the future, prefer to live in a world with many possibilities and options, are focused on complex abstract problems, and consider the complete picture, sometimes at the expense of details (intuition). Highly rated teachers accept other peoples' values and are good at estimating their influence on decision making (feeling). In addition, they prefer a flexible and adaptive life style, lean toward spontaneity, require more information when making decisions, and often do things at the last moment (perceiving). By contrast, the typical or average group of teachers displayed a different combination of types; they were dominated by sensing rather than intuition characteristics. They also relied on more sensible experiences, are oriented toward more practical problems, and believed that well-established routines should not be changed. And, finally, the judging type is more likely than the perceiving type to prefer order and to be self-disciplined and well-structured with an organized lifestyle.[4]

We offer here some problems that an extroverted teacher might want to use. However, we also recommend these problems for teachers who may consider themselves more introverted, especially if you tend to strictly follow the textbook and are looking for ways to move around the classroom and create a dynamic, fun learning experience for your students. Of course, as we mention throughout the book, selecting any material beyond or what is in the textbook is a decision the teacher must make based on the student population. Yet, if you are up to challenging your own self-perceptions about introversion—and if you tend to be more introverted—these problems might help you change that aspect of your personality, at least in the classroom

---

[4] I.B. Myers & P.B. Myers, *Gifts Differing: Understanding Personality Type*. Palo Alto, CA: Davies-Black, 1995.

setting. Essentially, by bringing an unusual problem or topic to the classroom, one that has motivated you in the first place, your changed behavior from the normal "following the curriculum" style will certainly be noticed by the class in a favorable fashion.

Here's the problem: *In a class experiment, Miguel rolls one ordinary six-sided die repeatedly, keeping track of each number he rolls. He decides to stop as soon as one number is rolled three times. Miguel stops after the twelfth roll, and the sum of these rolls is 47. Which number occurred for the third time? (An ordinary six-sided die has the numbers from 1 through 6 on its sides).*

One approach is to obtain a die and carry out the experiment. It will be difficult to get an exact sum of 47 in 12 rolls, but even if you did get the answer that would be the inelegant method!

Let's use our strategy of logical reasoning. After the eleventh roll, no number had yet appeared three times, otherwise the experiment would have already ended. This means that five of the numbers appeared twice, and the other number once. Let's call this number M. If M were rolled on the twelfth roll, the total would then have been $2(1+2+3+4+5+6) = 42$. Therefore, the total after 11 rolls is $42 - M$. If N is the number rolled for the third time, then $42 - M + N = 47$ and so $N - M = 5$. We know that N and M can only take on the numbers from 1 through 6. The only two numbers that allow a difference of 5 from these numbers are 6 and 1. Therefore, with this restriction the equation $N - M = 5$, has only one solution where $M = 1$ and $N = 6$. Thus, it is the number 6 which is rolled a third time.

We can consider a problem that would be perhaps of a more typical nature and can be adapted to the environment in which the class finds itself. Here is a problem that can be adapted to most school situations.

The problem is as follows: *At a school with 25 classes, each class sets up a basketball team to compete in a school-wide tournament. In this tournament a team that loses one game is immediately eliminated. The school only has one gymnasium, and the principal of the school would like to know how many games will be played in this gymnasium in order to get a winner.*

The typical solution to this problem could be to simulate the actual tournament by beginning with 12 randomly selected teams playing a second group of 12 teams–with one team drawing a bye—that is, passing up a game. This would then continue with the winning teams playing each other as shown here.

Any **12 teams** vs. any other **12 teams**, which leaves **12 winning teams** in the tournament.

6 winners vs. **6 other winners**, which leaves **6 winning teams** in tournament.

3 winners vs. **3 other winners**, which leaves **3 winning teams** in tournament.

3 winners + 1 team (which drew a bye) = **4 teams**.

2 remaining **teams** vs. **2** remaining **teams**, which leaves **2 winning teams** in the tournament.

1 team vs. **1 team** to get a **champion!**

Now counting up the number of games that have been played we get:

Teams playing	Games played	Winners
24	12	12
12	6	6
6	3	3
3 + 1 bye = 4	2	2
2	1	1

The total number of games played is: 12+6+3+2+1 = 24. This seems like a perfectly reasonable method of solution, and certainly a correct one—adopting a different point of view from what would have been expected.

Approaching this problem from a different point of view would be vastly easier by considering the losers, rather than winners, which is what we did in the previous solution. In that case, we ask ourselves, how many losers must there have been in this competition in order to get one champion? Clearly, there had to be 24 losers. To get 24 losers, there needed to be 24 games played. And with that the problem is solved. Looking at the problem from an alternative point of view is a curious approach that can be useful in a variety of contexts.

Another alternative point of view would be to consider these 25 teams with one of them—only for our purposes—considered to be a professional basketball team that would be guaranteed to win the tournament. Each of the remaining 24 teams would be playing the professional team only to lose. Once again, we see that 24 games are required to get a champion. This should demonstrate for you the power of this problem-solving technique. We now consider a wide variety of problems that can be most efficiently solved by adopting a different point of view.

Another problem that teachers might be able to adapt to a common situation is the following: *Each of the 10 court jewelers gave the king's advisor, Mr. Pogner, a stack of gold coins. Each stack contained 10 coins. The real coins weighed exactly 1 ounce each. However, one and only one stack contained "light" coins, each having had exactly 0.1 ounce of gold shaved off the edge. Mr. Pogner wishes to identify the crooked jeweler and the stack of light coins with just one single weighing on a scale. How can he do this?*

The traditional procedure is to begin by selecting one of the stacks at random and weighing it. This trial and error technique offers only a 1 chance in 10 of being correct. Once this is recognized, one may revert to attempt to solve the problem by reasoning. First of all, if all the coins were true, their total weight would be 10×10 or 100 ounces. Each of the 10 counterfeit coins is lighter, so there will be a deficiency of 10×0.1 or 1 ounce. But thinking in terms of the overall deficiency doesn't lead anywhere, since the 1-ounce shortage will occur whether the counterfeit coins are in the first stack, the second stack, the third stack, etc.

Let us try to solve the problem by organizing the data in a different fashion. We must find a method for varying the deficiency in a way that permits us to identify the stack from which the counterfeit coins are taken. Label the stacks #1, #2, #3, #4, ... #9, #10. Then we take one coin from stack #1, two coins from stack #2, three coins from stack #3, four coins from stack #4, etc. We now have a total of $1 + 2 + 3 + 4 + \cdots + 8 + 9 + 10 = 55$ coins. If they were all true, the total weight would be 55 ounces. If the deficiency were 0.5 ounces, then there were 5 light coins, taken from stack #5. If the deficiency were 0.7 ounces, then there were 7 light coins, taken from stack #7 and so on. Thus, Mr. Pogner could readily identify the stack of light coins, and consequently the jeweler who had shaved each coin.

What if you complete the MBTI questionnaire and are confronted with the qualities that define the average or typical teacher? Do not despair. The MBTI results do not have to be an iron-clad predictor of what type of teacher you will be. Instead, use the results as a starting point to think about your teaching style. The results should guide you to a new awareness that can help you improve as a teacher. As with any test of this type, there are criticisms of the MBTI's construction and questions about its reliability and accuracy in determining personality traits. This exercise should enable you to identify strengths and weaknesses so that you can make needed adjustments based on a new awareness. Once you have an idea of what your personality traits are as defined by the MBTI—and if you believe that analysis to be fairly accurate and useful—you can work toward changing or adjusting some of the things you may do in the classroom.

The Five-Factor Model can also be applied to a study of teacher personality to determine which traits are highly effective in the classroom. The factors in this model are meant to be understood as a continuum, so that, for example, a teacher who lands on the extreme end of the neuroticism factor would display the pronounced traits of negative emotions, pessimism, low tolerance for frustration, and impulsiveness.[5] Clearly these are not desirable traits for teachers who need to inspire trust and build dynamic relationships with their students. Research conducted using the Five-Factor Model to understand

---

[5] L. Goncz, Teacher Personality: A Review of Psychological Research and Guidelines for a More Comprehensive Theory in Educational Psychology. *Open Review of Educational Research*, 4(1), 75–95 (2017).

teacher personality supports the idea that extroversion is an effective trait for teachers—as long as it is moderate extroversion. Moderate extroversion is a trait associated with friendliness, self-confidence, and positive emotions. This mirrors the findings regarding extroversion in the MBTI research (in which the best teachers exhibited traits of extroversion). Extreme extroversion, on the other hand, might manifest itself as being too talkative and being unaware of others' desire to engage. We have all been exposed to this type of person: the long-winded lecturer, the show-off, the bore—certainly not effective traits for a teacher.

Remember that in addition to neuroticism and extraversion, the five factors also include openness to experience, agreeableness, and conscientiousness, and each factor operates on a continuum. Conscientiousness is a highly desirable trait in teachers, for example, but only if it is balanced (or moderate) and not taken to an extreme. You should cultivate a reasonable level of conscientiousness so that your characteristics of competence, order, planning, self-discipline, and impulse control do not devolve into rigidity and inflexibility. All five factors should be moderately balanced.

A look at the fields of journalism and theatre might shed some light on our exploration of ideal personality traits. In print journalism, for example, the qualities most often mentioned in connection with what makes a successful journalist include tenaciousness, curiosity, excellent writing and communication skills, courage, and open-mindedness. Many people might assume that a good journalist would also need to be an extrovert, since they often must pursue interview subjects who are resistant and/or reticent to talk. And once a journalist manages to land the interview, they must be able to draw out their subjects, get them to open up and reveal things that may be painful or may not be in their best interest. With this in mind, it would also be tempting to place a journalist on the higher end of the extroversion scale in the Five-Factor Model as well as in the three areas of openness to experience, agreeableness, and conscientiousness. However, as is often the case when dealing with all the complexities of human beings, there are those who are the exception to the rule and find ways to excel beyond the "normal" range. So perhaps it may not be entirely surprising to hear from successful journalists who not only describe themselves as introverts but also claim that their introversion contributes to their success as journalists.

Here, we offer an example from mathematics in which what appears to be intuitive or is assumed is actually incorrect (similar to the thinking about introverts as journalists). Sometimes what is obvious is wrong! We offer here a situation where cleverness can be shown mathematically. Suppose you had a job where you received a 10% raise. Because business was falling off, the boss was soon forced to give you a 10% cut in salary. Will you be back to your starting salary? The answer is a resounding (and very surprising) NO!

This little story is quite disconcerting, since one would expect that with the same percent increase and decrease you should be where you started from. This is intuitive thinking, but wrong. Convince yourself of this by choosing a specific amount of money and trying to follow the instructions. Begin with $100. Calculate a 10% increase on the $100 to get $110. Now take a 10% decrease of this $110 to get $99–$1 less than the beginning amount.

You may wonder where the result would have been different if we first calculated the 10% decrease and then the 10% increase. Using the same $100 basis, we first calculate a 10% decrease to get $90. Then the 10% increase yields $99, the same as before. So, order makes no difference.

A similar situation can be faced by a gambler, that is, one that is deceptively misleading. Consider the following situation. You may want to even simulate it with students to see if their intuition bears out. Here is how the game is played, with you as the player.

You are offered a chance to play a game. The rules are simple. There are 100 cards, face down. 55 of the cards say *win* and 45 of the cards say, *lose*. You begin with a bankroll of $10,000. You must bet one-half of your money on each card turned over, and you either win or lose that amount based on what the card says. At the end of the game, all cards have been turned over. How much money do you have at the end of the game?

The same principle as above applies here. It is obvious that you will win ten times more than you will lose, so it appears that you will end with more than $10,000. What is obvious is often wrong and this is a good example. Let's say that you win on the first card, you now have $15,000. Now you lose on the second card, you now have $7,500. If you had first lost and then won, you would still have $7,500. So, every time you win one and lose one, you lose one-fourth of your money. So you end up with $10,000 \times \left(\frac{3}{4}\right)^{45} \times \left(\frac{3}{2}\right)^{10}$. This is $1.38 when rounded off. Surprised?

Jen Retter, a journalist who considers herself an introvert (INFJ on the MBTI: introvert, intuition, feeling, judging) explained how she has used her introvert traits to connect with people and get them to open up: "Introverts are listeners. Being introverted doesn't mean you avoid people entirely 24/7, it just means you draw energy from solitude. Being around people is fine, so long as it's not all day every day and you have recharging time in between. When I'm in conversation, I listen. I read people well. I'm not caught up in trying to get my point of view in or draw attention to myself. I'm just listening, listening for the important piece of information to catch my attention. That's how I find stories from listening to random people sharing with me about whatever to latching onto a subject's sentence mid-interview and using that as my angle."[6] Retter has benefited from an awareness of her own introversion and from her ability to turn what may be perceived as an undesirable trait in journalism to her advantage. So, while extroversion is often assumed to be a desirable trait for a journalist, sometimes that obvious assumption is wrong.

Retter's insight into her own personality trait is a great example of how one can gain from an understanding of personality strengths and weaknesses. In teaching, an introvert can benefit from taking advantage of listening skills—an introvert's superpower!—while also managing to develop and put into practice a few characteristics from the extrovert type category. Just because you are an introvert does not

---

[6] J. Retter, How this Introvert Built a Successful Career as a Journalist. Retrieved on May 16, 2020 from: https://www.quietrev.com/how-this-introvert-built-a-successful-career-as-a-journalist/.

mean you cannot be a great teacher.[7] On the other hand, if you are an extrovert, make sure there is time and space in the classroom for students to participate. Try to moderate the extroversion. One of the most common complaints about the extreme extrovert is that they never "shut up."

One aspect of journalism that may also be enjoyable for the introvert is the time spent analyzing the interview and research material and then writing and re-writing the article. Writing is a solitary process and would be appealing to those with introversion traits. When it comes to the interview process, the introvert is able to keep the focus on the interviewee, which is a role the introvert would feel most comfortable taking. The journalist is not in the spotlight or part of the story, nor should he/she be. This dynamic resonates with certain situations in a classroom environment: the drawing out of a reticent student, or allowing a student to contribute an idea during a lecture or class activity. We can use the techniques of the journalist who focuses on the interview subject, asks a series of relevant and probing questions (and follow-up questions), while simultaneously analyzing the content that the interviewee offers and fact checking.

Just as journalists must motivate to get the most out of their subjects, mathematics teachers must motivate students to generate interest and maximize their learning receptivity. Mathematics can provide topics that can generate discussion and motivate students who are typically not as engaged as they might well be. Presenting a topic that is not part of the curriculum will surely spark an eagerness to investigate further. For example, consider some oddities in mathematics, which can open up the realm of mathematics to students in rather unique ways. One such oddity has to do with a number that seems to not want to disappear. This oddity is apparently a quirk of the base-10 number system and is referred to as the *Kaprekar constant*, which is the number 6,174. This constant arises when one takes a four-digit number and forms the largest and the smallest

---

[7] Learn more about the power of introverts in S. Cain, *Quiet: The Power of Introverts in a World that Can't Stop Talking.* New York, NY: Crown Publishers, 2012.

number from these digits, and then subtracts these two newly formed numbers. Continuously repeating this process with the resulting differences, will eventually result in the number 6,174. When the number 6,174 is reached, and the process is continued— that is, creating the largest and the smallest number, and then taking their difference (7,641 − 1,467 = 6,174), we will always get back to 6,174, which is the Kaprekar constant. To demonstrate this with an example, we will carry out this process with a randomly-selected number. When choosing the number, avoid numbers with four identical digits, such as 3333. Actually, the difference between largest and smallest digit must be at least 2. For numbers with less than four digits, you obtain four digits by padding the number with zeros on the left, such as 0012. For our example, we will choose the number 3023:

- The largest number formed with these digits is: 3320.
- The smallest number formed with these digits is: 0233.
- The difference is: 3087.
- The largest number formed with these digits is: 8730.
- The smallest number formed with these digits is: 0378.
- The difference is: 8352.
- The largest number formed with these digits is: 8532.
- The smallest number formed with these digits is: 2358.
- The difference is: 6174.
- The largest number formed with these digits is: 7641.
- The smallest number formed with these digits is: 1467.
- The difference is: 6174.

And so, the loop is formed, since you will continue to get the number 6174.

In the case where you choose a four-digit number, where the largest digit differs from the smallest digit by less than two, the result of this "Kaprekar process" will be zero. In all other cases, you will always end up with the number 6,174, which then gets you into an endless loop (i.e. continuously getting back to 6,174). It should never take more than seven subtractions. If it does, then there must have been a calculating error. Incidentally, another curious property of 6,174 is that it is divisible by the sum of its digits:

$$\frac{6174}{6+1+7+4} = \frac{6174}{18} = 343.$$

Once again, benefiting from a journalist's experience, we note that he/she must also be willing to meet with and talk to people of various backgrounds and experiences with opinions that may not align with their own values and beliefs. Journalists must be open to these situations. They must also be flexible in how they approach their subjects in order to secure interviews. In other words, they must know what motivates people. And, in line with another of the Five-Factor Model's categories, journalists must be conscientious in their work. They must verify the stories of their interviewees and fact check each article they write or else they will gain a reputation as unethical and/ or unreliable. While there are some interesting points of connection between journalism and teaching, especially in reference to the Five-Factor Model, we can also look to techniques used in theatre and drama to explore personality.

There are many different approaches to acting: the Meisner technique, Stanislavski's system, and Lee Strasberg's method, among others. Each approach requires actors to delve deeply into their emotions and their memories or to use their imagination to experience the emotions of the character they are portraying. Whether they are accessing their own emotional reserves or are creating emotions based solely on character, actors practice, and sometimes perfect, the techniques of becoming someone else. When an actor successfully achieves this transformation, on stage or on film, viewers are willing to suspend their disbelief,[8] and they enter into a world of

---

[8] A theatrical reference that refers to an intentional avoidance of critical thinking or logic in examining something surreal (work of fiction or an operatic performance, for example) in order to believe it for the sake of enjoyment.

entertainment and art. This dramatic process of becoming someone else may be instructive to those who want to better understand their own personality traits and who may be searching for ways to modify some of those traits. Exploring emotions and memories may be useful in any attempt to modify personality traits that are more conducive to the classroom.

We do not have to become actors to become better teachers, but we might want to employ some of the techniques from the actor's repertoire. Two of those techniques include improvisation and role-playing. Improvisation is not stand-up comedy, although there may sometimes be elements of improv in stand-up comedy. Rather, improvisation is performed without any preparation or a script. Through the practice of improvisation, we must think on our feet, so to speak, and to react immediately. Improvisation does not allow us to over-analyze a situation, but it does force us to listen patiently. The focus in improvisation is on the process—the process of interacting in real time with another human and accepting the unexpected. Some people find that engaging in improvisation heightens their awareness and liberates them from some of their fears and anxieties. This process has the potential to reveal unexamined personality traits and can also be an opportunity to practice behaviors and thought processes that might be useful in changing certain personality traits. In addition to the potential for teachers, other industries beyond theater have started to use improvisation in professional development exercises to help employees and their companies reach their goals.

Business schools and boardrooms, for example, have hired the well-known Chicago-based improv troupe *Second City* to train employees in improvisation in the belief that it promotes entrepreneurship, nurtures creativity, and builds leadership skills.[9] They are also hoping that employees and students who take improv classes improve their team-building skills, communicate insights more effectively,

---

[9] S. White. How an Improv Class Can Help Develop Essential Business Skills. *Financial Management*, 2/1/2018. Retrieved on May 15, 2020 from: https://www.fm-magazine.com/issues/2018/feb/improv-class-helps-develop-business-skills.html.

understand their own body language and read the body language of others,[10] improve their listening skills, and prepare for the unexpected. Another technique that businesses borrow from theater to use in management and sales training is role-playing. Role-playing can be similar to improvisation; however, there is a script of sorts in role-playing. It is an active learning technique in which employees act out situations under the guidance of a trainer who provides feedback to the participants. Role-playing exercises can be valuable in the business world when participants learn how to experience a situation from another perspective, learn how to respond to customer complaints, or to practice how to be more persuasive in sales situations. Role playing for teachers can provide opportunities to question self-perceptions regarding personality traits, can provide insight into behaviors, and can be a first step in the "fake it until you make it" strategy that psychologist Carol Dweck suggests for those who may want to adjust a personality trait.

Can we really change our personality or is our basic personality fixed for life? The Austrian psychoanalyst Sigmund Freud believed that our personality was, for the most part, pretty much established by the age of five.[11] Modern psychologists seem to support Freud's claim.[12] If we are shy introverts who struggle with public speaking, can we ever manage to be an excellent teacher? If we are extroverts who draw energy from interactions with others and crave the spotlight, can we ever learn to turn the spotlight on the students or guest speakers in our classroom? In this chapter, we do not suggest that a major personality change is possible or even advisable. What we posit is that with self-awareness and some practice, we can alter certain aspects of our personality traits to become more effective teachers in the classroom.

---

[10] The importance of body language is covered in Chapter 3 "Know How to Reach Your Students."

[11] G. Frank, Freud's Concept of the Superego: Review and Assessment. *Psychoanalytic Psychology*, 16(3), 448–463 (1999). https://doi.org/10.1037/0736-9735.16.3.448.

[12] R. Mõttus, W. Johnson, & I. J. Deary, Personality Traits in Old Age: Measurement and Rank-Order Stability and Some Mean-Level Change. *Psychology and Aging*, 27(1), 243–249 (2012). https://doi.org/10.1037/a0023690.

Some psychologists suggest that "broad" personality traits are fairly stable throughout our lives, but it is the "in-between" spaces of our personality that may be more mutable. One of those psychologists is Carol Dweck, who describes those in-between spaces of personality as beliefs and belief systems and goal and coping strategies. Dweck claims that changing the behavior patterns, habits, and beliefs that lie under the surface of broad traits such as introversion or agreeableness, is the key to personality change.[13] She suggests focusing on the following four areas:

(1) **Change habits:** Habits can be changed. It is not easy, but with practice, new patterns of behavior can emerge and start to feel natural.

(2) **Change self-beliefs:** Charles Horton Cooley's concept of the "looking-glass self" suggests that we develop our sense of self through the perception and judgment of others.[14] Changing our self-beliefs can be challenging, especially if the judgment and perception of others have had too much influence on our self-perception. However, just as developing new habits is possible, it is possible to change those aspects of our personality that do not serve us in our role as educator.

(3) **Focus on process:** Dweck outlines the difference between a "growth mindset" and a "fixed mindset" regarding personality.[15] Focusing on process supports a growth mindset in that it puts an emphasis on effort rather than ability. Remember, improvisation helps the participant to focus on the process.

(4) **Fake it until you make it:** Psychologist Christopher Peterson recounts how he felt that his introverted personality might have a negative impact on his career as an academic. Peterson's strategy

---

[13] C. S. Dweck. Can Personality Be Changed? The Role of Beliefs in Personality and Change. *Current Directions in Psychological Science*, 17(6), 391–394 (2008). https://doi.org/10.1111/j.1467-8721.2008.00612.x.

[14] C. H. Cooley. *Human Nature and Social Order*. New York, NY: Charles Scribner's Sons, 1922.

[15] C. Dweck. *Mindset: The New Psychology of Success*. New York, NY: Ballentine Books, 2007.

for change illustrates how the technique mentioned in the discussion on drama can be effective. To overcome his introversion, he decided to start "acting" extroverted in certain situations, for example, when delivering a lecture to a class full of students or giving a presentation at a conference. Over time, and through habit, these behaviors simply became second-nature. While he acknowledged that he was still an introvert, he learned how to become extroverted *when he needed to be.*[16]

---

[16] K. McGowan, Second Nature. *Psychology Today*, 3/1/08, updated 6/9/16.

Chapter 2

Know Your Audience

Anyone who has been asked to deliver a speech realizes the importance of knowing who the audience will be and, perhaps equally as important, the purpose of the speech. As the speech-giver, have you been asked to inform, to motivate, to entertain, or all three? What are the particular *needs* of the audience? Research from the field of communication studies and best practices in media industries emphasize the need to consider other factors as well such as geographical and cultural concerns, current events and context, the size and age of the audience, appropriate appearance and tone (formal or informal, for instance), and audience knowledge.

When educators meet a new classroom full of students for the first time, they are already juggling a plethora of requirements set forth by the state, the city, and their own school district. But one of the most basic—and important—questions they will need to ask and answer is: "Who is my audience?" If educators can answer this question, they are on their way to making meaningful connections to students through the material they use, the assignments they design, and group projects they create that are actually relevant to their students. Relevancy is key to encouraging students to care and is key to unlocking a student's love of learning.

Now that you understand who you are as an educator and how your personality actually informs your teaching style (Chapter 1), the focus can shift to your audience. A number of other fields and industries such as media and business recognize the importance of understanding and connecting with an audience. Educators can borrow some of the best practices from other fields to reach students. In fact, educators and students would benefit from the integration of some of these best practices into commonly used and effective teaching approaches. Many experts in the field of education point to four basic cognitive issues that educators should understand: students' prior knowledge, how that knowledge is organized, how knowledge is accessed, and how students' monitor, evaluate, and adjust their learning strategies (also known as metacognition). This chapter focuses on the first item in this list—students' prior knowledge—as it relates to getting to know and understand an audience.

Understanding Students' Prior Knowledge: Audience Knowledge

What is the current standard of measurement for students' prior knowledge? They are sitting in your classroom because they "passed" the previous grade. There were a number of challenges that each student had to successfully overcome and they had to demonstrate proficiency in specific areas of study. Should you assume, however, that just because a group of students passed into the next level of their education that they can work at grade level in your classroom? Probably not. Yet many teachers do not find out whether their students are at grade level until that first assignment, quiz, or test—and, unfortunately, many have no idea who their students are in a more general sense.

Educators would greatly benefit from having a variety of innovative methods by which to get to know their students *quickly*. While getting a sense of students' grade level is critical, it is imperative that educators also gain a more comprehensive view of their students as whole human beings—not only as the "student" as determined by grade level, achievements, test scores, and so forth, but as a whole person who experiences a physical, emotional, psychological, and spiritual world and is already shaped by their place in that world.

Understanding who your students are from the start will ensure that you chose material that will resonate with them and with their personal living experiences. In addition, when educators know what their students know, they can build upon that knowledge and strengthen students' reception for new information.

Educators should strive to learn about the various geographical, social, and cultural aspects of their students' lives. For example, are you teaching in an urban, suburban or a rural area? Each will present its own challenges; however, each will present wonderful opportunities to find relevant material for use in the classroom and will also enable the design of assignments that will resonant with your "audience." Show a child who has always lived in a rural area the opening scene of the popular children's television program "Sesame Street" and they might be confused about the urban street scene of brownstones and sidewalks. A home to this group does not look like one brownstone attached to another brownstone; and a landscape that includes sidewalks and small patches of dirt with one tree is completely foreign. This group of students, for the most part, has grown up in wide open spaces with no sidewalks and in detached houses. On the other hand, use a math problem with language that includes a reference to a "lawn" in a classroom filled with students who live in high-rise apartment buildings and you will most likely get a high-level of confusion. Yet, with a very basic level of knowledge about students, educators can consider ways to incorporate various landmarks from the students' community into material and assignments or incorporate current events from that community into material and assignments. Once recognizable landmarks and or current events from a shared community become the focus of an assignment, the level of student interest increases. The material becomes much more relevant to the student, and relevancy encourages students to care.

We can all agree that it is a math teacher's responsibility to pay particular attention to problem-solving skills as an integral part of every instructional program. Suppose you are involved with a secondary school classroom in a large urban setting with a subway system where some trains are express trains skipping local stations and other grains stop at every station. Let's take the New York City subway system as an example. Suppose you are at the 14th Street station and the No. 1 train and the No. 2 train are both accepting passenger simultaneously. Your objective is to go to the 137th Street station as quickly as possible. This is a local stop only reachable by the No. 1 train. The No. 2 train is an express train and skips many stations, but can only be taken to the 96th Street station running parallel to the No. 1 train stopping along the way at a few stations at which the No. 2 train also stops, as it then veers off in another direction away from the 137th Street station. The question you are faced with is which train are you better off boarding? Clearly, if you board the No. 1 train it will get you to your destination. On the other hand, the better strategy would be to take the No. 2 train since it might overtake an earlier local No. 1 train and get you to the 137th Street station earlier than the other No.1 train you might have boarded at 14th Street. This is a very useful problem-solving strategy in mathematics called "considering extreme situations." Or as we sometimes say in everyday language "a worst-case scenario," since if the No. 2 train does not overtake an earlier No. 1 train you can still get on the original No. 1 train you chose not to board originally.

Let's consider a mathematics problem that can be more easily solved by considering extreme situations. Once this concept has been introduced using a student's home environment—the subway—and the student gains confidence in their grasp of the concept, they can go on to practice this concept using other problems not directly related to their "prior knowledge."

In a drawer, there are eight blue socks, six green socks, and 12 black socks. What is the smallest number of socks which must be taken from the drawer, without looking at the socks, to be certain of having two socks of the same color?

At first glance this problem appears to be similar to the model problem discussed previously. In this case, we are looking for a matching pair of socks of *any* color. We now apply *extreme case* reasoning. The worst-case scenario has us picking one blue sock, one green sock, and one black sock in our first three picks. Thus, the fourth sock must provide us with a matching pair, regardless of what color it is. The smallest number of socks to guarantee a matching pair is 4. Even though our two examples are different in nature, the commonality is that we used in extreme situation to help us solve the problem.

We can also use a geometry problem that would be rather eye-opening for many students, as it is typically not presented in this more sophisticated fashion in the usual classroom setting. Consider the problem of proving the following theorem[1]:

> From a point in an equilateral triangle the sum of the distances to the three sides of the triangle is a constant.

There are several ways to prove this theorem. However, before embarking on a proof, it would be desirable to know what this "constant" is. We can inspect this question by considering extreme positions for this randomly placed point. Suppose we place this point on a side of the triangle (despite its violation of the given condition, namely, being *in* the triangle). This would have the effect of reducing the distance to this side to 0. How could we reduce the distances to two of the sides of the triangle to 0? By placing the point so that it lies on two sides at the same time, that is, at a vertex. Now a revisit to the original question becomes trivial. The sum of the distances to the three sides is now simply the altitude of the equilateral triangle. So, we have found out what this constant probably is: the altitude of the triangle. Therefore, the use of examining an extreme case, was an important aid in the pursuit of a solution. For once we know more precisely what we have to prove, the more easily we should be able to accomplish the task.

Being cognizant of the particular social and/or cultural "norms" in your region is also of enormous benefit to educators, and, as a result, to students. Ideas about privacy, modesty, food, physical space, and contact, or even ideas about wearing shoes indoors or eye contact, among many other "norms," vary widely from region to region. It is incumbent upon the educator to know how these factors

---

[1] For a few proofs of this theorem as well as some interesting applications of the theorem, see A.S. Posamentier, *Advanced Euclidean Geometry*. Emeryville, CA, Key College Publishing, 2002.

might manifest in the classroom. And while it may seem impossible to *completely* understand all the ways these factors might impact your classroom, especially if you are new to the community where you teach, your students will appreciate your effort. Of course, the longer you stay at your particular school, you will begin to learn more about your student population, or your student "demographics."

When businesses set out to understand their target market, for instance, they often look to statistical data—or demographics—in order to identify and then segment the population that is more likely to be receptive to their products or services. When businesses understand the needs of their target market, they are in a better position to address those needs. If a business fails to address the needs of their target market, the business will undoubtedly fail. In much the same way, when educators are familiar with the demographics of their own student population, they can become more effective in the classroom.

Demographics are statistical data that usually consist of categories such as race, ethnicity, gender, age, education, profession, income level, marital status, and number of children or size of family. Which of these categories apply to your particular audience, of course, will depend on the age range of your students. At the very least, educators should consider the race, ethnicity, gender (and sexual orientation—more about this in Chapter 10), and the income level of the student population in their classroom. Again, this information makes it possible for educators to use relevant material and to design interesting assignments. Remember, the main goal here is to get students to care, and using material that is relevant to their lives draws students in, it piques their interest and encourages them to care. When educators are innovative in the ways that they add a layer to existing assignments or content or in the ways they create a new assignment with new content, students are much more likely to be motivated to understand concepts. That additional layer—or new and relevant content—should be based on appropriate demographic information that facilitates a connection between student and content. When students feel connected, they will care. An important point: The demographic information gathered is not intended to target a particular student or group of students but rather to build a "classroom profile" that enables educators to motivate and engage students. Introducing new material or content should, of course, not be based on stereotypes of certain groups. Educators must demonstrate a grasp of the current and appropriate language and always use sensitivity when discussing group or individual identity (Chapter 10 will address this in more detail).

Typically, during the course of mathematics instruction, teachers usually credit male mathematicians such as Pythagoras, Euclid, and even the United States President James A. Garfield, who is credited with an original proof of the Pythagorean theorem. However, rarely do they mention that women also have played significant roles in the development of mathematics as we have it today. For example, in 1843, the British mathematician Ada Lovelace produced what we consider today the first foray into computer programming, by embellishing on the *Difference Machine* developed by the British mathematician Charles Babbage. Such inclusions of female contributors to the history of mathematics would go a long way to "include" all students in the class and perhaps serve as a more general motivator.

We also have a tendency to base history on European experiences rather than expanded to other cultures beyond Europe. Research over the past century has shown that the famous theorem accredited to Pythagoras, was known well before his time. It is entirely possible that Pythagoras might have seen this relationship on other occasions, as it was well known in other parts of the world long before Pythagoras. In Mesopotamia, mathematicians were even able to produce further triples of numbers fulfilling the Pythagorean condition $a^2 + b^2 + c^2$, as we can see on a Babylonian clay tablet from ca. 1800 BCE, known as the Plimpton 322[2] (see Figure 2.1). The tablet was written in the sexagesimal system (base 60) using the cuneiform script of the times. It shows us the high level of mathematics knowledge that existed well before the Greeks.

---

[2] This tablet is in the permanent collection of the Columbia University (New York) Library.

Figure 2.1

The transcript of these cuneiforms as shown in Table 2.1, provides some Pythagorean triples.

Here, we note that the three left-numbers in each line satisfy the Pythagorean theorem, $a^2 + b^2 + c^2$, and are called Pythagorean triples, while the two numbers on the right side (labeled m and n) are the numbers that can be used to generate these triples, as follows: $a = 2mn$, $b = m^2 - n^2$, $c = m^2 + n^2$.

Utilizing current events can also be a way to draw students into the subject matter, especially if those current events speak to students' experiences and interests. The news cycle moves quickly presenting many opportunities for teachers in terms of variety of content. With a world of media at our fingertips—thanks to the Internet—teachers can access stories about people and events from Baghdad to Bangor. There are also wonderful resources for teachers that collate news stories into grade levels, so when choosing a story to use in class teachers can be confident that it is grade appropriate (one good

Table 2.1

a	b	c	M	n
120	119	169	12	5
3456	3367	4825	64	27
4800	4601	6649	75	32
13500	12709	18541	125	54
72	65	97	9	4
360	319	481	20	9
2700	2291	3541	54	25
960	799	1249	32	15
600	481	769	25	12
6480	4961	8161	81	40
60	45	75	2	1
2400	1679	229	48	25
240	161	289	15	8
2700	1771	3229	50	27
90	56	106	9	5

source is *Newsela*, a service that provides standards aligned content). Carefully selecting current events that highlight people and topics that students can relate to, however, depends entirely on the teachers' knowledge of their students. If teachers do not know their students, choosing current event topics will be hit or miss and waste valuable class time. Even worse, students will lose interest. At its best, using current events in the classroom can provide the added benefit of introducing or emphasizing the importance of media literacy. With an overabundance of information constantly available to students through digital and print media, it is critical that they develop the ability to recognize legitimate news sources, the ability to recognize an opinion and a fact as well as develop the ability to use multiple news sources to get a more comprehensive view of any story or event.

What are some good examples of using current events creatively in the classroom? Whether teaching math or science, government, economics, culture, the arts, literature or history, there are some best practices for matching current events to content areas. Try creating an exercise in which students analyze data side by side or in which they compare news sources on an event to see how various sources treat the news differently. Students might also create an infographic based on a particular news story or take the data from a news story and create a graph or chart. Conversely, create an exercise that asks students to write a story about a data set that describes what that data "says." All of these exercises help students understand how to translate information—from visual to narrative, from raw data to narrative, or from narrative to visual.

One problem that students can work on in groups would be the following. Have them consider that most folks find percentage problems to have long been a nemesis. Problems get particularly unpleasant when multiple percents need to be processed in the same situation. This student activity can turn this one-time nemesis into a delightfully simple arithmetic algorithm that affords lots of useful applications and provides new insight into successive percentage problems. This not-very-well-known procedure should enchant students. Let's begin by considering the following problem:

> *Wanting to buy a coat, Barbara is faced with a dilemma. Two competing stores next to each other carry the same brand coat with the same list price, but with two different discount offers. Store A offers a 10% discount year-round on all its goods, but on this particular day offers an additional 20% on top of their already discounted price. Store B simply offers a discount of 30% on that day in order to stay competitive. How many percentage points difference is there between the two options open to Barbara?*

At first glance, they may assume there is no difference in price, since $10 + 20 = 30$, yielding the same discount in both cases. Yet with a little more thought they may realize that this is not correct, since in store A only 10% is calculated on the original list price, with the 20% calculated on the lower price, while at store B, the entire 30% is calculated on the original price. Now, the question to be answered is, what percentage difference is there between the discount in store A and store B? One expected procedure might be to assume the cost of the coat to be $100, calculate the 10% discount yielding a $90 price, and an additional 20% of the $90 price (or $18) will bring the price down to $72. In store B, the 30% discount on $100 would bring the price down to $70, giving a discount difference of $2, which in this case is 2%. This procedure, although correct and not too difficult, is a bit cumbersome and does not always allow a full insight into the situation.

An interesting and quite unusual procedure can be provided for entertainment as well as a fresh look into this problem situation: Here is a mechanical method for obtaining a single percentage discount (or increase) equivalent to two (or more) successive discounts (or increases).

(1) Change each of the percents involved into decimal form:

0.20 and 0.10

(2) Subtract each of these decimals from 1.00:

0.80 and 0.90 (for an increase, add to 1.00)

(3) Multiply these differences:

$(0.80)(0.90) = 0.72$

(4) Subtract this number (i.e. 0.72) from 1.00:

$1.00 - 0.72 = 0.28$, which represents the combined *discount*.

(If the result of step 3 is greater than 1.00, subtract 1.00 from it to obtain the percent of *increase*.)

When we convert 0.28 back to percent form, we obtain 28%, the equivalent of successive discounts of 20% and 10%.

This combined percentage of 28% differs from 30% by 2%.

Following the same procedure, students can also combine more than two successive discounts. In addition, successive increases, combined or not combined with a discount, can also be accommodated in this procedure by adding the decimal equivalent of the increase to 1.00, where the discount was subtracted from 1.00 and then continue the procedure in the same way. If the end result turns out to be greater than 1.00, then this end result reflects an overall increase rather than the discount as was found in the above problem. This procedure not only streamlines a typically cumbersome situation, but also provides some insight into the overall picture. For example, the question "Is it advantageous to the buyer in the above problem to receive a 20% discount and then a 10% discount, or the reverse: 10% discount and then a 20% discount?" The answer to this question is not immediately intuitively obvious. Yet, since the procedure just presented shows that the calculation is merely multiplication, a commutative operation, we find immediately that there is no difference between the two.

So, here they have a delightful algorithm for combining successive discounts or increases or combinations of these. Not only is it useful, but it also gives you some newly-found power in dealing with percentages, when using a calculator is not appropriate.

Current events can also be used to illustrate about how a particular lesson might operate or be reflected in the "real" world—or in the world outside of the classroom: There are entertainments in mathematics that stretch (gently, of course) the mind in a very pleasant and satisfying way. This example presents just such a situation. It requires some "out of the box" thinking that might leave your students with some favorable lasting effects. Let's consider the question:

*Where on earth can you be so that you can walk **one mile south**, then **one mile east**, and then **one mile north** and end up at the starting point?*

Mostly through trial and error a clever student will stumble on the right answer: the North Pole. To test this answer, try starting from the North Pole and traveling south one mile and then east one mile. This takes you along a latitudinal line which remains equidistant from the North Pole, one mile from it. Then travel one mile north to get you back to where you began, the North Pole. Most people familiar with this problem feel a sense of completion. Yet we can ask: Are there other such starting points, where we can take the same three "walks" and end up at the starting point? The answer, surprising enough for most people, is *yes*.

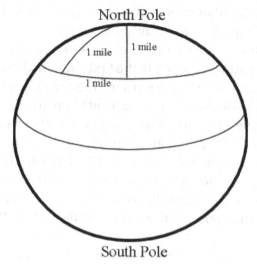

(Not drawn to scale, obviously!)

One set of starting points is found by locating the latitudinal circle, which has a circumference of one mile and is nearest the South Pole. From this circle, walk one mile north (along a great circle,

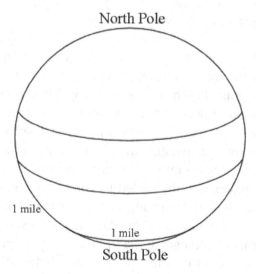

(Not drawn to scale, obviously!)

naturally), and form another latitudinal circle. Any point along this second latitudinal circle will qualify. Let's try it.

Begin on this second latitudinal circle (the one farther north). Walk one mile south (takes you to the first latitudinal circle), then one mile east (takes you exactly once around the circle), and then one mile north (takes you back to the starting point). Suppose the first latitudinal circle, the one we would walk along, would have a circumference of $\frac{1}{2}$ mile. We could still satisfy the given instructions, yet this time walking around the circle *twice*, and get back to our original starting point. If the first latitudinal circle had a circumference of $\frac{1}{4}$ mile, then we would merely have to walk around this circle *four* times to get back to the starting point of this circle and then go north one mile to the original starting point.

At this point we can take a giant leap to a generalization that will lead us to many more points that satisfy the original stipulations, actually an infinite number of points! This set of points can be located by beginning with the latitudinal circle, located nearest the south pole, which has a $\frac{1}{n}$ th-mile circumference, so that a 1-mile walk east (which comprises n circumnavigations) will take you back to the point on the circle at which you began your walk on this latitudinal circle. The rest is the same as before, that is walking one mile south and then later one mile north. Is this possible with latitude circle routes near the North Pole? Yes, of course!

Knowing your students will also help guide when, where, and how you use a formal or informal teaching approach in presenting material and in your day-to-day interactions. What is the difference between formal and informal teaching and learning? A formal classroom environment is generally very structured, lessons and discussions are led by the teacher, and, in this environment, learning takes place *in the classroom*. An informal classroom environment is generally more interactive, not always led by the teacher, sometimes it is student- or group-led, and learning sometimes takes place outside of the classroom—in a museum or on a field trip, for example. Informal teaching and learning recognize the influence of peer groups, media, and family as sources of information and knowledge. When teachers have a good understanding of their students, they will know when to use a more formal approach and when to use an informal approach to a particular lesson plan. Knowing your students will guide your ability to identify or develop leaders and to encourage more participation from reserved students through the use of both formal and informal approaches to assignments. For instance, if a teacher has decided to teach a collaborative approach to problem solving, the teacher might create groups and assign a problem for group members to solve together. The teacher might ask that one student in particular represent the group and report the findings.

There are lots of problems that groups of students can hover around to try to solve. The example that we give here would be one that would be more appropriate at the secondary school level and involves having the group prove that there are exactly five regular polyhedra that exist. One possible approach would be to consider the following. In Euclid's book, *Elements* he also describes their construction and indicates that there are no other regular polyhedra beyond the tetrahedron, cube, octahedron, dodecahedron, and icosahedron. There are number of ways in which the students can prove that there are only these five-regular polyhedral. The proof that Euclid uses in *Elements* goes as follows: The solids are comprised of vertices, faces and edges and each has its limits. The vertices are comprised of at least three faces. However, the plain angle of each face at a vertex must be less than 120°, since if they were each 120°, then the sum of 3 such would be 360° and there would be no vertex, just a plane.

As we look at regular polygons, those with 6 or more sides have angles exceeding 120°, and therefore, are not eligible to be faces of a regular polyhedron. Thus, we will consider faces that are regular polygons of 3, 4, and 5 sides, namely, triangular faces, square faces, and pentagonal faces. The regular triangle has vertices of 60°. Placing them to form a three-dimensional vertex, commonly known as a polyhedral angle, we find that there are three possibilities and they generate the tetrahedron, octahedron, and icosahedron. When we consider faces of the square, where the vertex angle is 90° the only possibility for creating a polyhedral angle with face angles of 90° is that which forms a cube. Now considering a possible face regular polygon which is a pentagon, each vertex angle is 108° which again allows us only three faces at the polyhedral vertex leading us to construct a dodecahedron. When we count the number of regular polyhedra thus constructed we end up with 5.

At the middle school level, students might be motivated to consider some elementary probability through the famous "birthday problem." They might begin by visiting 10 classrooms and in each having students indicate their birth date (without the year) on a small slip of paper. The group can then inspect the 10 piles of papers to see how many of them have a match of birthdates. This will lead to a formal investigation of the expected results, which are truly counterintuitive. This activity presents one of the most surprising results in mathematics. It is one of the best ways to convince the uninitiated in the "power" of probability. The results of this activity, aside from being entertaining, will upset the students' sense of intuition.

To better understand this situation, let's consider the following example. Suppose you are in a class with about 35 students. What do you think the chances (or probability) are of two classmates having the same birth date (month and day, only). Intuitively one usually begins to think about the likelihood of two people having the same date out of a selection of 365 days (assuming no leap year). Perhaps 2 out of 365? That would be a probability of $\frac{2}{365} = .005479 \approx \frac{1}{2}\%$. A minuscule chance.

Let's consider the "randomly" selected group of the first 35 presidents of the United States. You may be astonished that there are two with the same birth date:

the 11th president, James K. Polk (November 2, 1795), and

the 29th president, Warren G. Harding (November 2, 1865).

You may be surprised to learn that for a group of 35, the probability that two members will have the same birth date is greater than 8 out of 10, or $\frac{8}{10} = 80\%$.

If you have the opportunity you may wish to try your own experiment by selecting 10 groups of about 35 members to check on date matches. For groups of 30, the probability that there will be a match is greater than 7 out of 10, or in 7 of these 10 rooms there ought to be a match of birth dates. What causes this incredible and unanticipated result? Can this be true? It seems to go against our intuition. To satisfy your curiosity we will consider the situation in detail.

Let's consider a class of 35 students. What do you think is the probability that one selected student matches his own birth date? Clearly *certainty*, or 1.

This can be written as

$$\frac{365}{365}.$$

The probability that another student does *not* match the first student is

$$\frac{365-1}{365} = \frac{364}{365}.$$

The probability that a third student does *not* match the first and second students is

$$\frac{365-2}{365} = \frac{363}{365}.$$

The probability of all 35 students *not* having the same birth date is the product of these probabilities:

$$p = \frac{365}{365} \cdot \frac{365-1}{365} \cdot \frac{365-2}{365} \ldots \frac{365-34}{365}.$$

Since the probability (q) that two students in the group *have* the same birth date and the probability (p) that two students in the group do *not* have the same birth date is a certainty, the sum of those probabilities must be 1. Thus, $p + q = 1$.

It's In this case,

$$q = 1 - \frac{365}{365} \cdot \frac{365-1}{365} \cdot \frac{365-2}{365} \ldots \frac{365-33}{365} \cdot \frac{365-34}{365} \approx .8143832388747152.$$

In other words, the probability that there will be a birth date match in a randomly selected group of 35 people is somewhat greater than $\frac{8}{10}$. This is quite unexpected when one considers there were 365 dates from which to choose. The motivated student may want to investigate the nature of the probability function. Here are a few values to serve as a guide:

Number of people in group	Probability of a birth date match
10	0.1169481777110776
15	0.2529013197636863
20	0.4114383835805799
25	0.5686997039694639
30	0.7063162427192686
35	0.8143832388747152
40	0.891231809817949
45	0.9409758994657749
50	0.9703735795779884
55	0.9862622888164461
60	0.994122660865348
65	0.9976831073124921
70	0.9991595759651571

Notice how quickly almost-certainty is reached. With about 60 students in a room the chart indicates that it is almost certain (0.99) that two students will have the same birth date.

Were one to do this with the death dates of the first 35 presidents, one would notice that two died on March 8th (Millard Fillmore in 1874 and William H. Taft in 1930) and three presidents died on July 4th (John Adams and Thomas Jefferson in 1826, and James Monroe in 1831). Above all, this astonishing demonstration should serve as an eye-opener about the inadvisability of relying too much on intuition.

The difference between formal and informal can also refer to teaching style. Some teachers prefer a more disciplined, highly-structured, and focused approach to teaching while others might prefer a more relaxed and friendly approach—with a goal of creating a rapport with students and building trust. Knowing yourself and how your personality affects your ideas about who you are as a teacher (Chapter 1) is critical because in order to take advantage of both informal and formal strategies in teaching, you will need to know your strengths, weaknesses, and proclivities when it comes to teaching. You will need to reflect and perhaps adjust your own ideas and beliefs about yourself as a teacher. What you want to avoid is reverting to a "default" setting in which you are acting without fully understanding who you are and why you might choose one method over another, for instance. When you are in the default setting, you are not acting with intention and self-understanding. Figure out if you have a predisposition toward a more formal approach or tend toward a more informal approach. Consider experimenting with different teaching methods, and try to strike a balance between the two approaches so that you can adjust to what works best for your students.

Information Collection Methodologies

Now that you fully appreciate the benefits of knowing your audience—of knowing your students—how do you go about collecting the information you need? There are a few methods to consider. First, if possible, talk with your colleagues, especially the ones who just spent a year teaching the students who are entering your classroom environment. Make sure to maintain a sense of objectivity when listening to colleagues' accounts, and carefully choose which colleagues with whom you speak. The critical balance in this data collecting is to glean useful information about students while rejecting information that is based on rumor, bias, or some other non-relevant personal viewpoint. Talking to other colleagues could also be useful to determine if your incoming class of students have siblings in higher or lower grades at your school as well as other information about special family dynamics, past challenges, and/or talents and accomplishments.

Other methods that may prove useful in getting to know students include using short surveys/questionnaires, short-answer writing exercises, conducting focus groups, or asking students to interview each other and recording the interview to share with the class. These methods are valuable tools to learn about students in your class and should be used as early in the school year as possible. Teachers can create their own tools or use a variety of free resources widely available to educators. The type of questions you include on the survey or questionnaire will depend on what you want to know about your students, however, the most common categories are generally student goals, family and home environment, personal information, attitudes towards school, educational background, and writing prompts that may be specific to your class/subject area.

A first day of school "get to know you" survey can include questions that will reveal feelings about school (Describe school using one word), feelings about teachers (Who was your favorite teacher and why?), feelings about expectations and/or ability (What grade do you expect to get in this class? Do you think you are smart? Why/why not?), and may provide the student with the opportunity to tell the teacher how he or she can be supportive (What is the most important thing I can do to help you succeed?). Teachers may also want to include questions about extracurricular activities, favorite things to do outside of school, and the perennial "what did you do over the summer break?" Depending on the grade level of the class, teachers might also consider a short-answer worksheet that gives students the chance to write about their favorite things: favorite subject, favorite sport/activity, favorite lunch, favorite book, etc. Students can also answer questions that reveal levels of confidence or self-esteem such as: "In school, I'm proud that I_____" or "I can help others with_____."

In addition to these methods, teachers can create focus groups that address specific topics. Focus groups can be conducted not only at the beginning of the school year but at any point during the year as a way to learn more about the students' experiences, including feedback on subject material, teaching approaches, learning goals, assessment, support needed, etc. Focus groups can be a fun, low-stakes way to prompt discussion about student likes and dislikes as well as what areas they may need help with or find boring or too easy. Marketers and advertisers often use focus groups to get a sense of the viability of a product—to observe group participants' reactions to a product and to facilitate a discussion about the product (or service). In all focus groups—whether it is with students in your classroom or participants in a product-oriented focus group—to collect accurate information, the facilitator (or teacher) must remain neutral and objective. There are, of course, no wrong answers and participants should be encouraged to be honest and forthright. It is also common to provide some kind of incentive to focus group participants, and in the case of students, refreshments or extra time doing something they enjoy might be appropriate. The ideal size of a focus group is 7–10 participants, and teachers should rotate the students chosen to participate so that each student has had a chance to provide feedback. If the technology is available, teachers may decide to videotape the focus group discussion so they can review later to catch comments they may not have heard initially or taken down in notes.

All of these methods should be undertaken as early in the school year as possible and can be introduced as fun "getting to know you" exercises. Teachers might even ask students to exchange surveys or questionnaires as a starting point to interviewing each other or having a more in-depth discussion. Depending on the grade level and the technology available, teachers may decide to have students record or videotape their survey/questionnaire responses or interviews on a mobile device. There are a variety of free apps available for this type of exercise. However, the majority of the methods mentioned in this section do not require technology and can also be accomplished simply with paper and pencils and conversation.

Chapter 3

Know How to Reach Your Students

In business, when marketers put together a strategy to introduce and sell a product, they address the "Four Ps." The Four Ps are that standard marketing mix of product, price, promotion, and place. This chapter focuses on "place" (or distribution channel), which dictates, in part, how businesses *reach their target market*. It is part of a larger strategy to determine how the product will be delivered to the customer or how the customer will access the product. This idea of "place" in marketing can translate to "communication" in the classroom and the ways in which we communicate with our students, and also being aware of how students communicate with each other—something that is often encouraged by teachers.

The distribution channel is the initial place where the customer is engaged. What distribution channels are best for your target market? As educators, we should ask ourselves, what channels or techniques are available to us to reach our students? More ideally, what are the *best* channels to utilize to reach your current set of students most effectively? Once you get to know your audience (recall Chapter 2, "Know Your Audience"), you will be able to identify the best channels to use to reach your students. If you don't fully understand your student "target market," you won't know which channels are the best to use.

Since we are concerned with teaching students, not marketing to potential customers, we need to understand what "place" or distribution looks like in the world of education. What are these channels? At the very basic level, distribution channels can be best understood as how we communicate with our students both verbally and nonverbally. In order to foster trust, understanding, and effective two-way communication, teachers need to develop good communication skills. While some might say: "Well, that's just common sense!"—unfortunately, *common sense* is not necessarily *common practice*.

The commonly accepted definition of communication is the act of transmitting and/or expressing ideas, information, knowledge, thoughts, and feelings as well as understanding what is expressed by others. Communication is the act of sending and receiving messages. It might seem simple and straightforward, but we all know that communication is anything but that! We have all been in the position of thinking that we communicated an idea or concept clearly, but when confronted by the look of confusion on the faces of our students (or friend or colleague), we realize that our message may not been effectively transmitted.

There is a myriad of factors that come into play when we communicate with another person. Do we share the same cultural values and beliefs? Is the person I am talking to distracted, fearful, bored, etc.? Am I speaking loud enough? What is my tone of voice? How well do they understand the language in which I am speaking? And much, much more. Also consider that communication occurs one-on-one, in a group setting, in a written or printed format, or visually—all these different scenarios represent additional layers of complications.

As educators, we also need to be aware of the two basic forms of communication: verbal and nonverbal. How much attention do we pay to the many forms of nonverbal communication? After all, it is not just *what* we say, but *how* we say it. Nonverbal communication includes tone of voice; smiling, frowning, eye rolling and other facial expressions; body language such as crossed arms, eye contact, or lack of eye contact, hand gestures and other types of movement. Communication experts estimate that between 60% and 90% of the meaning of a message is conveyed nonverbally. If this is the case, it is easy to understand how much meaning can be lost through communicating by phone, email, or social media. Also consider how much communication is altered or even completely missed if a student is visually challenged or distracted in the classroom setting. On the other hand, imagine how much potential there is to effectively communicate with your students in a positive, encouraging, and motivating way through nonverbal communication cues and methods.

Perhaps it is best to start with understanding how your students communicate and expect to be communicated with in the classroom environment. What cues are they expecting from you? How can you best read students' nonverbal communication cues? This, of course, depends on age and grade level. In business, for example, marketers would want to know whether their particular target market members prefer to make purchases online or through a mail order catalog? Or do they prefer an in-store experience when shopping? These entail very different forms of communication. Online, mail-order catalogs, and in-store are all channels of distribution, yet each one requires a different marketing approach. Depending on the grade level, communicating with your students in the classroom might be best achieved through a variety of methods specific to that particular group of students.

The first step in determining the best channel(s) of communication is to analyze the information you gathered on your students through previous performances, interviews, surveys, questionnaires, and other methods covered in Chapter 2. The second step is to adapt this analysis to your own personal style of teaching or adapt your style of teaching when and where necessary to your target group of students. In this chapter, we will also consider strategies and best practices from a few fields outside of education. Effectively reaching your students is largely about effective communication, both verbal and nonverbal. Here we will explore some best practices mainly in nonverbal communication from the fields of law and sports coaching that may be adapted to your own classroom environment.

Eye Contact

When words are restrained, the eyes often talk a great deal.
— Samuel Richardson

Trial lawyers are generally known for their command of language. They need to master the nuances and subtleties of verbal communication to finesse an argument and to persuade juries and judges on behalf of their clients. For the trial lawyer—the lawyer who argues cases in the courtroom in front of a judge and/or a jury—a sophisticated understanding of nonverbal communication is also critical to successfully presenting an opening statement or a closing argument for their cases as well as examining witnesses. One of the key forms of nonverbal cues for a lawyer is effective eye contact. Stories about the importance of eye contact are prevalent in Western culture: It is said that the eyes are the window to the soul. There is also the old adage, "Don't trust someone who won't look you in the eye." In most Western cultures, making direct eye contact communicates trustworthiness and openness; eye contact (direct or lack of) is a form of nonverbal communication.

For trial lawyers and teachers, there are three reasons to look your audience/students in the eye: (1) Eye contact helps your students trust you. Conversely, they might not have as much trust in you if you don't make eye contact—just as you are probably less trusting of someone who won't meet your gaze. (2) Eye contact helps you connect with the students. Have you ever been in an audience and the speaker made eye contact with you? Did that eye contact make you pay more attention? Did it make you feel like the speaker was interested in you? Maybe you even felt more of a connection after that eye contact was made. Litigators know that even casual eye contact can have a tremendous effect on jurors. (3) Eye contact helps you "read" your students. Some trial lawyers bury their heads in their notes and never look at the jurors. Big mistake! The jurors could be bored, confused, even falling asleep and the lawyer would completely miss those signs. If that were the case, the lawyer would not have the opportunity to make adjustments in what and how they are communicating verbally. If you are correctly reading those nonverbal cues, you might get the chance to slow down, speed up, ask a question, or change course entirely. In this sense, reading nonverbal communication cues ensures that teachers (and trial lawyers) have the option to adjust their verbal communication, which allows them to express their ideas more effectively.

A few other points about the importance of eye contact from the courtroom are worth considering. For instance, when using visual aids, avoid looking at the visual aid to the detriment of making eye contact with your students. Trial lawyers using visual aids, such as photographs or flip charts or other types of evidence, must make a concerted effort to keep eye contact with the jury. While the jurors are looking at the visual, the lawyer needs to assess the jurors' reactions, just as a teacher needs to assess students' reactions to visual aids. Focusing on your students' nonverbal cues will make you a more effective teacher, and as you gain experience, you will be able to look out into the sea of students sitting in your classroom and register interest, boredom, confusion, excitement, or, hopefully, real engagement.

One last point about eye contact. Make an effort to connect with *each* one of your students through eye contact. If you ignore one of them, they will notice! Jump from student to student, holding eye contact with each one for a few seconds—so that it seems like a natural conversation between the two of you. You might be tempted to make eye contact more frequently with the students who are engaged and who tend to speak out on a regular basis in class. This, however, would be a mistake. It sends a powerful message to those students who are not engaged that you are not connected to them—you are not drawing them into the "circle of involvement." Eye contact is powerful. Use it to convey sincerity, to share a moment of levity, or to draw students back into your lesson and into a conversation.

There are some very attractive topics beyond the curriculum to which students may eventually gravitate. When presenting a topic of this sort, it is important for teachers to keep their eyes on the class to see how they react and to modify the presentation accordingly. Let's consider one such somewhat unusual topic that is often used by teachers at a variety of grade levels, namely, the Fibonacci numbers.[1] This series of numbers, which many consider the most ubiquitous numbers in all of mathematics, stemmed from a problem of the regeneration of rabbits in a book by Leonardo of Pisa, who later was referred to as Fibonacci. These numbers are: 1, 1, 2, 3, 5, 8, 13, 21, 34, 55, 89, 144, . . . , where each succeeding number is the sum of its 2 predecessors. An example of where a teacher would easily be distracted from making eye contact with the students in the class would be where the teacher would like to demonstrate that even on a typical pineapple, we can find the Fibonacci numbers represented. The hexagonal bracts on a pineapple can be seen to form three different-direction spirals. In Figures 3.1–3.4, we notice that in the three directions there are 5, 8, and 13 spirals. These just happen to be three consecutive Fibonacci numbers.

Figure 3.1

[1] For more about the Fibonacci numbers, see: *The Fabulous Fibonacci Numbers*, by A. S. Posamentier, and I . Lehmann, Guilford, CT: Prometheus Books, 2007.

Figure 3.2

Figure 3.3

Figure 3.4

When showing the class this incredible fact and holding the pine-apple in your hand while perhaps marking the three kinds of spirals with the chalk marker, it is very likely that a teacher will take his or her eyes off the class and concentrate solely on the pineapple. Here, a teacher must make every effort to keep his or her attention on the class members and not on the pineapple. If need be, have a student hold the pineapple and a few students mark the three spirals.

An ambitious teacher may wish to take this a step further, since this can also be done with pinecones. There are various species of pinecones (e.g. Norway spruce; Douglas fir or spruce; Larch). Most will have two distinct direction spirals. Spiral arrangements are classified by the number of visible spirals (parastichies) that they exhibit. The number of spirals in each direction will be most often two successive Fibonacci numbers. The two pictures of the Norway spruce pinecones shown in Figure 3.5 will bear this out, but you may want to convince yourself by getting some actual pine cones and counting the spirals yourself.

These pinecones have eight spirals in one direction and 13 in the other direction. Again, we will notice these are Fibonacci numbers.

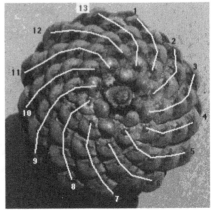

Figure 3.5

Tree (species)	Number of spirals in right direction	Number of spirals in left direction
Norway spruce	13	8
Douglas fir or spruce	3	5
Larch	5	3
Pine	5	8

Once again, the effective teacher will concentrate on making eye contact with the class and not be distracted by the objects being presented. Granted, this can be a bit difficult, however, an effective teacher will keep this eye contact factor in mind.

Physical Proximity

Where you stand depends on where you sit.

— Nelson Mandela

Anthropologist Edward T. Hall, who was an expert on nonverbal communication, described four zones of space that exist around a person: (1) *intimate* space, extending out about eighteen inches; (2) *personal* space, stretching out from one and a half feet to four feet; (3) *social* distance, reaching out to twelve feet; and (4) *public* distance in the space beyond.[2] Hall's theory of space had a significant impact on communication theory, especially intercultural communication theory. Space between people can represent how friendly or close they are when having a conversation, and the appropriate amount of space varies by culture. Different cultures have different rules as to how close someone can be when having a conversation—the space between people can reflect a social setting as well as their cultural background.

Hall described *social* distance as the space used by people who work together, while people who are at a public distance are outside the "circle of involvement." But what does this have to do with the

---

[2] Edward T. Hall, *The Hidden Dimension*. Garden City, NY: Doubleday, 1966.

field of law? Just as a person's use of eye contact can influence how he or she is perceived, a lawyer and client's relative proximity to the jury box also has influence. If the lawyer and client sit within social distance (4–12 feet) from the jury box, research shows they might have an advantage in making a personal connection with jurors over the opposing party sitting farther away.[3] In other words, a closer proximity offers a distinct advantage since it gives the impression of being within what Hall refers to as the "circle of involvement."[4] Social distance is within the circle of involvement—and if lawyers can occupy this space they have the opportunity to communicate certain nonverbal cues and gain a perceived connection with the jurors. While this may not be the distance reserved for family and friends (personal space), it is the distance usually reserved for acquaintances (seating arrangements in the courtroom are often within the trial judge's discretion).

Now apply this concept of space to the classroom environment. The "circle of involvement" seems an appropriate phrase when thinking of a classroom setting. Teachers want to create an environment in which every one of their students feels a part of the group and involved. If that is to be achieved, the teacher may want to stay out of the public distance area (beyond 12 feet) and within the social distance (4–12 feet). When teachers get into the public distance realm, they may become less effective. They should choose carefully when and how long they will stand at the front of the class. This does not mean that teachers should never be more than twelve feet away from the students. Rather they should strive to strike a balance. Use common sense as a common practice.

There are other ways that teachers can use Hall's theory of space to their advantage in the classroom. For example, unless there is assigned seating, the most engaged students may often choose to sit in the front of the class. This can also be interpreted in a

---

[3] Jeffrey S. Wolfe, The Effect of Location in Courtroom on Jury Perception of Lawyer Performance. *Pepperdine Law Review*, 731, 769–771 (1994).

[4] Edward T. Hall, *The Hidden Dimension*. Garden City, NY: Doubleday, 1994, pp. 108–122.

classroom arranged in groups of students rather than in the traditional rows. Students in the back of the classroom perhaps are hoping that there is an invisible zone in the last few rows of the room or a portion of the room where they are hidden from the teacher's gaze. Of course, there are always some exceptions to this self-selected seating strategy. In general, however, students not wishing to engage sit in the back and students who want to engage or want to be noticed, sit in the front part of the classroom—in the circle of involvement. With an understanding of the social space (within twelve feet), the teacher can flip the front of the classroom to the back just by walking to the back of the room and delivering announcements and/or lessons in the social space of those students in the back of the room. You will get their attention immediately. Once you have their attention, make eye contact and invite them to participate. Use this strategy of moving around the classroom often to encourage more engagement from students in that "invisible zone" in the back of the room. Draw your students into the circle of involvement!

Where a teacher stands in the classroom can also affect the degree to which students can follow the material being presented. For example, if a teacher wants to demonstrate how one can fold a strip of paper into a knot to form a regular Pentagon and then hold it up to the light so that the diagonals can be noticed, as we see in Figure 3.6, naturally, those students in closest proximity will gain the most. Teachers need to be aware of their location with respect to the entire class.

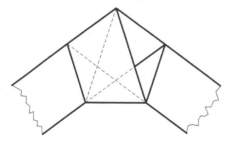

Figure 3.6

Another example of where a student's proximity to the teacher makes a difference is when a teacher wants to use the following technique to show that the sum of the measures of the exterior angles of any polygon is 360°. Here, you will see how proximity is an essential part of this lesson and requires the teacher to perform accordingly.

Begin the lesson by showing students Figure 3.7 and ask them to focus on the sum of the measures of the exterior angles. Either have a very large version of this figure redrawn on a large piece of paper or the board or have a series of drawings of this figure each significantly smaller than its predecessor (see Figures 3.8–3.10). Figure 3.10 should be so small or distant from the students so that the polygon almost appears as a point. Then have students conjecture about the sum of the measures of the exterior angles, which now may look like they are lines emanating from a point rather than extensions of sides of a polygon.

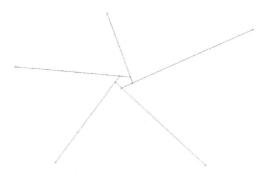

Figure 3.7

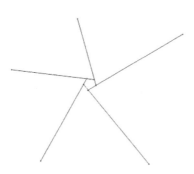

Figure 3.8

Figure 3.9

Figure 3.10

The effect of reducing this figure successively can be achieved either by moving the figure further away from the class or by presenting them with progressively smaller figures. Students should eventually realize that the sum of the measures of the exterior angles appears as a complete revolution about a point (the extreme situation, where the polygon reduces to a point) or 360°. Again, here the distance from students is an essential part of the presentation.

Remember that students constantly observe your nonverbal cues, so when you move to the back of the classroom those students may gain a clearer sense of what you are communicating. You are now actually close enough for them to better see your hand gestures and facial expressions, as well as hear your tone of voice, and other, more subtle, nonverbal communication cues. The importance of understanding what you are communicating to your students through nonverbal communication cues is important for a number of reasons. Your nonverbal cues should be in sync with your verbal communication in order to reinforce the point you want to make. You do not want to be in the position of communicating one thing with your words and another with your facial expression. If you roll your eyes, for example, while making a point during a lesson, students may interpret that nonverbal cue as "this material is a joke!" Teachers are the authority figure in the classroom, and your students will pick up on biases that you may be unknowingly communicating through nonverbal cues.

In the legal system, trial judges—the figure of authority and power in the court—are obligated to maintain a façade of impartiality. The legal system recognizes the influence of judges and has instituted certain practices and procedures to limit an appearance of bias, including simplifying instructions to the jury so that they are less likely to rely on nonverbal cues from the judge. Excessive interventions by a trial judge, after all, diminishes the effectiveness of the judicial system. This example is relevant to nonverbal communication in the classroom environment because of something called the "Rosenthal Effect." Named for Robert Rosenthal, the psychology professor who studied this phenomenon, the Rosenthal Effect occurs when individuals modify their behavior to conform with what they perceive to be the expectation of the person in authority. If adults alter their behavior to conform with those of a person in a position of power, imagine young children's susceptibility to the teacher's influence. While we hope to model positive behavior in the classroom, we also might unwittingly be communicating other messages through nonverbal cues. These messages might not even be negative messages, just counter to what you think you are conveying or are attempting to convey. Just like trial judges, teachers are human beings, not robots or automatons. We are not required to be a great stone face that shows no reaction to anything that happens in the classroom. However, we must have a good sense of what we are communicating nonverbally.

We also might take a cue from sports coaching to better understand our nonverbal communication habits. In the world of sports, coaches commonly video record practices and games and then view them afterwards to analyze the athletes' performances. These video recordings are played over and over, pausing them or putting them into slow motion in order to more closely examine an error, to identify patterns, or reconfigure a play. Video can be used in a similar way as a part of a teacher's training and assessment. With a high-quality recording device in practically all cell phones, it is possible to record teaching sessions at a relatively low cost. If teachers are able to view themselves teaching, they can identify their nonverbal communication cues—both positive and negative. Such a viewing assessment could be done with or without a supervisor depending on the particular procedures of the teaching program. Ultimately, the goal would be to identify the nonverbal communication cues, gain an awareness of how those cues may or may not affect students, and then work to eliminate the negative cues.

On the positive side, a teacher often shows unusual enthusiasm when presenting a topic that can enrich the class and yet may not be part of the standard curriculum. For example, at a particular point in time, where it might be advantageous to demonstrate that division by zero is not permitted, the teacher can approach the class and say "would you like me to prove to you that $1 = 2$?" The teacher's enthusiasm here can be clearly noticed visually.

A convincing case for defining a way division by zero is not permitted is to show students how it can lead to a contradiction of an accepted fact, namely, that $1 \neq 2$. You can show them that were division by zero acceptable, then $1 = 2$; clearly an absurdity!

Here is the "proof" that $1 = 2$:

$$\text{Let } a = b$$
$$\text{Then } a^2 = ab \qquad \text{[multiplying both sides by } a\text{]}$$
$$a^2 - b^2 = ab - b^2 \qquad \text{[subtracting } b^2 \text{ from both sides]}$$
$$(a-b)(a+b) = b(a-b) \qquad \text{[Factoring]}$$
$$a + b = b \qquad \text{[Dividing by } (a-b)\text{]}$$
$$2b = b \qquad \text{[replace } a \text{ by } b\text{]}$$
$$2 = 1 \qquad \text{[Divide both sides by } b\text{]}$$

In the step where we divided both sides of the equation by $(a - b)$, we actually divided by zero, because $a = b$, so $a - b = 0$. That ultimately led us to an absurd result, leaving us with no option other than to prohibit division by zero. By taking the time to explain this rule about division by zero with your students, they will have a much better appreciation for mathematics. Remember, that the teacher's disposition during this activity will be quite different from that which typically can be seen during the normal instructional program.

By viewing yourself in an observational video, you might notice that when a student is answering a question in class you squint your eyes or burrow your forehead. You might recognize this cue as something you "naturally" do when you concentrate. To a student, however, your burrowed brow may appear as a grimace signaling displeasure. Your facial expressions may not match what you are actually thinking or feeling, and students are sensitive to that. Remember the Rosenthal Effect and how your position of authority can impact students' behavior in unintended ways.

Also, observe how you walk around the classroom (or not!) and use space as defined by Edward Hall. Are you in the "circle of involvement"? Or do you generally stand in one place at the front of the classroom? Social distance (4–12 feet) is ideal to create a sense of involvement, and perhaps you might think about organizing small group settings for your students that fall into Hall's category of personal space (18 inches to 4 feet). This space might encourage a closer, team-building environment. Also observe your body language. Do you use hand gestures to make a point or are you a bit wooden and lacking in other types of body language? Being overly theatrical can be a distraction, but a comfortable and confident style of movement can be a positive model for students. Observe the ways that students are using the space in the classroom and think creatively when setting up the class. Using space can empower a lesson plan as well as inspire and support the bonding of your students.

This chapter has focused mainly on nonverbal communication in the classroom. Using examples of best practices from the fields of law and sports reveals and underscores the power of nonverbal communication cues. In the courtroom, it can mean the difference between a successfully argued case or guilty verdict. In sports coaching, analyses of athletes' performances through video is a practice that can be used in teaching training and assessment, especially as it relates to a teacher's own nonverbal communication cues. Watching your own performance in the classroom can be an eye-opening experience. There is much to be gained from an insight into your own facial expressions, tone and pitch of voice, and body language.

Chapter 4

Know What Motivates
Your Students

People often say motivation doesn't last. Well, neither does bathing—
that's why we recommend it daily.

—Zig Ziglar

Promoters know what motivates their target audience. Good managers know how to motivate their employees. Once educators learn how to effectively communicate with their students, they can employ the strategies and techniques in this chapter to motivate them to learn. While implementing these various tools will increase student motivation, teachers also can borrow some best practices from the field of management to motivate students. Good managers know the importance of understanding what motivates their employees—they strive to create the right workplace atmosphere and to generate positive emotions to get the best performances from their direct reports. Interestingly, the approach to motivation in management has just as much to do with the supervisor's leadership skills and style as it does with the employees' level of motivation.

This chapter explores how management and leadership best practices can transfer to the classroom environment and impact your students' level of motivation. In addition, this chapter will present concrete examples of techniques that have been proven to motivate students to learn. Understanding what motivates your students will allow you to help your students embrace a lifetime love of learning. By communicating with your students about what in particular motivates them, you will be putting them in a better position to succeed in many areas of their life, from personal relationships to college to career.

What do we mean when we talk about motivation? A common definition of motivation is "the reason or reasons one has for acting or behaving in a particular way; it is the general desire or willingness of someone to do something." It all boils down to "wanting."[1] To understand motivation is to understand what causes a certain behavior and why that behavior might vary in its intensity. It is to understand, too, what is behind the "wanting." What causes a person to exercise, for example? Some people exercise to achieve a personal goal or perhaps for general health benefits. A group of high school students might exercise because their coach told them that as members of a football team, practice is mandatory. Some people exercise for fun or enjoyment, while others are focused on practicing a particular move to achieve competence—as gymnasts do. Sometimes the motivation is driven by more than just one factor. You might have noticed in these examples that for some the motivation is internally driven, while for others, such as the football players, the motivation is from an external source. This leads us to examine two main sources of motivation.

Motivation can be driven from extrinsic and intrinsic sources. Mandatory football practice is an external source—the football coach who requires participation in practice. In order to be on the football team, students know they have to show up and perform. Their reward is to remain on the team, and they avoid the punishment of being thrown off the team by attending practice. Or perhaps the reward is the praise they receive from performing well in practice. Yet, the football player may also be motivated by an intrinsic source as well. Intrinsic sources of motivation come from within. They can be task-related (a desire to understand), ego-related (a desire to perform), or

---

[1] Lexico/Oxford: http://www.lexico.com/en/definition/motivate, R.C. Beck, *Motivation: Theories and Principles*, 5th edn. Englewood Cliffs, NJ: Prentice Hall, 2004.

social-related (the desire to impress others). If teachers can discover the sources of motivation that already exist within their students, half the battle is won. Once these sources of motivation in each student are revealed, teachers can also nurture the appreciation for other sources of motivation in their students. For example, a teacher could help a student who is motivated to outperform others recognize the value in learning to understand—as opposed to the source of motivation being *solely* to outperform others. In addition to using the knowledge of motivation sources, and depending on the particular student population, teachers might choose to help students recognize what motivates them and bring this knowledge to a higher level of consciousness. In addition to the other more tangible techniques offered in this chapter, this self-knowledge has the potential to empower students in all aspects of their lives.

Motivating students is not a one-way street—it is a complex dynamic involving various traits of both the teacher and the student. While the focus on what motivates students is critical, teachers must also learn how to motivate students using their *own personal style* of teaching (Chapter 1). This is where educators can adapt some best practices from the field of business management and leadership to the classroom setting. In the world of business, *management* is defined as the organizing and the setting of strategies of a business through the coordination of efforts of employees to accomplish the objectives of the business. This is done through the application of available financial, natural, technical, and human resources. The four main functions of management are planning, organizing, leading, and controlling.[2] In this definition, some characteristics of the classroom environment and the relationship between teacher (manager) and students (employees) become apparent.

Relatively recently, there has been a greater focus on the development of leaders and leadership skills in business. This is not exactly surprising since research has shown that effective leadership in any business generally has the positive outcome of a stronger brand

---

[2] M. Alvesson, T. Bridgman, & H. Willmott (eds.), *The Oxford Handbook of Critical Management Studies.* Oxford: Oxford University Press, 2009.

identity, and, ultimately, increased profits.[3] In the classroom setting, a teacher with strong leadership skills can motivate students to achieve their own personal goals as well as the goals put forth by the school district, which often times emanate originally from the state education department.

Educators motivate students through their own leadership in the classroom, and this is when the focus on motivation turns to the teacher's leadership style and skills. It can be instructive to analyze aspects of the world of business—mainly management and leadership—to explore what makes an effective leader and what qualities of a leader motivate a group of employees. People in business and in academia are invested in understanding whether a leader is born or made. Researchers who study leadership have found that to be an inspiring leader, one does not need to have a special gift or the extraordinary charisma that is often assumed to be necessary or associated with effective leadership. Current research shows that leadership skills can be developed—one does not necessarily have to be born with the charisma of a leader. Through their data-driven approach, behavioral statisticians Joseph Folkman and John Zenger identified six characteristics that are most often associated with a strong leader: visionary; enhancing; driver; principled; enthusiast; and expert.[4] Sometimes the leader possesses more than one of these qualities, however, Folkman and Zenger's research found that possessing even one of these six characteristics can be enough to be an effective leader. What might these six characteristics look like in a classroom setting?

The *visionary* is the leader who provides a clear picture of the future, someone who clearly and consistently communicates what is expected, for example, in an assignment or a project or for what is expected in terms of behavior. The visionary encourages trust through

---

[3] J. Folkman & J. Zenger, *The Inspiring Leader: Unlocking the Secrets of How Extraordinary Leaders Motivate*. New York, NY: McGraw-Hill, 2009.

[4] J. Folkman & J. Zenger, *The Extraordinary Leader: Turning Good Managers into Great Leaders*, New York, NY: McGraw-Hill, 2002. Folkman and Zenger found that the last two characteristics (expert and enthusiast) were the least important out of the six.

this consistency—students can rely on clear instructions and clear expectations. The *enhancer* is the leader who is good at developing positive relationships, the enhancer makes an effort to connect one-on-one with students and is a good listener. The enhancer is that teacher who makes each student feel special through eye contact, personal interaction, and detailed knowledge about each student (because they listen carefully). In popular psychological terminology, the enhancer would be said to have a highly-developed "social intelligence."

Have you ever had a teacher or a manager who is focused on setting and achieving goals, and who motivates others to participate in reaching those goals? That person might take on the role of a cheer-leader: "One more client to recruit before we hit our goal of 100!" or "Fifty percent of the class has completed the task, let's help the other fifty percent reach their goal!" The *driver* is the leader who is focused on setting and achieving goals and excels in the pursuit of account-ability. If a teacher has the driver leadership style, it does not mean that they are solely focused on reaching goals and accountability—it just means that they excel in motivating others to set and achieve goals. This can be a powerful asset in education since so much of what a teacher must manage includes standards and assessment: setting goals, measuring outcomes, and overall accountability in the class-room for each students' learning. A focus on accountability can take different forms. For example, the driver can motivate students to work together to achieve goals, teaching important lessons in collabo-ration. If the driver provides students with the resources necessary to achieve goals, students will experience the positive emotions that propel motivation.

The *principled* leader is the role model who manages to set a posi-tive example that motivates others to follow. Again, consistency in behavior is key since it engenders trust among the students and between the students and teacher. Students trust the principled leader because they have observed the teacher's fairness when deal-ing with even the smallest matters. The principled leader inspires and motivates students to do their best through their own example as the classroom role model. The *enthusiast* leader has a passion and energy for the subject that inspires the students to learn. After creating a safe classroom environment in which students feel comfortable (actually all teachers should strive to do this no matter their leadership style), this enthusiasm can manifest in a number of ways: By creating drama in the classroom—the good kind, and using humor in an appropriate way. Enthusiasts are animated in both tone of voice and body language, and they find ways to use suspense and surprise in their lessons. An enthusiast might also bring a pantograph to the class to demonstrate the property of similarity. Using props, such as a pineapple or pinecone to illustrate the Fibonacci numbers concept, is also a good strategy. An enthusiast might also use the classroom as a creative space thorough vibrant decoration and student artwork.

The final leadership characteristic is the *expert*. The expert leader has a strong technical direction that comes from a deep expertise. The role of the expert in the business setting does not neatly translate to the classroom environment because there are clearly moments when being the expert in business brings a value ultimately measured in profits. Furthermore, unlike a business setting, there is little doubt among students that the teacher is an expert in his or her field. There is the curious situation, however, in which a teacher accidentally makes an error on a chart or on the board. Rather than to say, "I am only human and make mistakes as well," the teacher can cleverly escape the error by asking the students if the equation (or whatever is being demonstrated) is correct. If students are able to point out the error, the teacher can then simply say: "I made the error intentionally to see if you would catch it." Of course, the teacher cannot do this too frequently, but for the occasional faux pas this would be a good way of maintaining the "expertise" and putting some humor into the instruction program.

Each one of these leader types can build trust with students through positive relationships, good judgement, and consistency. Which one of these leader types best describes you? Remember that you might actually possess more than one of these characteristics. Zenger and Folkman interviewed and surveyed thousands of employees to create these six leadership characteristics. The list reflects the employees' responses to questions about motivation in the workplace and how managers have motivated them. If you are starting to feel some pressure to be that perfect leader, remember that all great leaders have weaknesses! This is not about perfection in the classroom. Rather it is about knowing who you are as a teacher, where your strengths lie, and how to cultivate and develop those strengths to motivate your students. Great leaders do a few things well—they do not excel in all areas. Leadership manifests as the presence of strengths, not the absence of weaknesses.

Ten Techniques to Motivate Your Students[5]

Over the past decade, brain imaging, data gathering methods, and behavioral science experiments have provided insights into the conditions that help us work more effectively. Business managers have tapped into this new information to support the development of strong leaders, which ultimately impacts the financial strength of the business. While the six characteristics of leadership reviewed in this chapter should prompt teachers to consider how they might lead their students to academic success, there are also some concrete techniques that have been proven successful. Combining the two—leadership motivation style with these techniques—will help inspire students to do their best work.

The following list of ten motivational techniques is a conceptual guide. Teachers can use the concept from each example and adjust to accommodate their own particular group of students or individual students. The examples should be used with the understanding that some stories and/or situations may not make sense to all students based on their own cultural beliefs and values. Once you get to know your students (Chapter 2), you will be able to adjust these techniques accordingly.

---

[5] A.S. Posamentier & S. Krulik, *Effective Techniques to Motivate Mathematics Instruction*, 2nd edn. New York: Routledge, 2016.

(1) Indicate a gap in students' knowledge—and help them close it!

Most students have a natural curiosity that can lead to a desire to complete their knowledge of a topic. This technique builds upon that desire by pointing out a void in the student's knowledge; this motivational technique works particularly well with any topic that has various components and where the last component to be presented would indicate completion and conclusion. First, present a few simple exercises involving a familiar situation. Students will have mastered this portion of the topic and feel a level of confidence in finishing these exercises. Second, present unfamiliar situations within that same topic and demonstrate to the class how the topic presented will complete their knowledge on that particular topic. This technique can be quite effective especially if it is presented in a dramatic fashion—a teacher's enthusiasm can be put to good use here.

Examples

Topic: Introducing the quadratic formula

The following is one example of how we can use this technique, which allows students to recognize that there is something they don't know yet but would like to know to complete their knowledge of a particular topic. The students have already learned how to solve quadratic equations presented in various formats by factoring. This activity will motivate them to recognize the need for a method for solving those quadratic equations when the quadratic polynomial is not factorable.

The teacher can begin the lesson by giving the students the following equations to solve. They will likely realize that the first three equations can be solved by factoring, but the last one cannot be solved by factoring.

Solve for x:

(1) $x^2 + x - 6 = 0$ $\quad$ $[(x+3)(x-2) = 0; x = -3, 2]$

(2) $x^2 - 25 = 0$ $\quad$ $[(x+5)(x-5) = 0; x = +5, -5]$

(3) $2x^2 - 5x - 3 = 0$ $[(2x + 1)(x - 3) = 0; x = -1/2, 3]$

(4) $x^2 - 9x + 7 = 0$ [unfactorable over whole numbers]

The students should feel comfortable solving the first three equations, since each can be solved by a method with which they are familiar, namely factoring. Students should recognize that this last equation is not solvable by factoring over the whole numbers. This should lead them to ask, if you are going to teach them how to solve this one in today's lesson. They have then perceived a void in their knowledge of solving unfactorable quadratic equations. The quadratic formula will resolve their dilemma. Students are now receptive, and hopefully eager, to learn the formula.

At some point at the start of the lesson it would be wise to indicate to the class (after they have learned the quadratic formula) that this formula can be used to solve *all* quadratic equations, even those that *can* be solved by factoring. To drive this point home you may wish to go back and solve each of the first three equations on their worksheet using the newly developed formula. Then using the newly developed formula they can then, finally, solve the fourth equation that frustrated them earlier in the lesson.

Topic: Introducing Heron's formula to find the area of a triangle

One of the more amazing formulas in geometry is Heron's formula, which allows one to find the area of a triangle without knowing an altitude's length and just having the length of each of the three sides of the triangle. At this point students are familiar with the usual formula for finding the area of a triangle is $Area = \frac{1}{2}bh$, where b is the length of the base and h is the length of the altitude to that side of the triangle. In geometry, there are other formulas for the area of triangles that students may already have learned: for a right triangle the area is one-half the product of its lengths, and to find the area of a triangle, where the lengths of two sides and the measure of the included angle are given, the area is one-half the product of the lengths of the two given sides and the sine of the included angle (i.e. $Area = \frac{1}{2}ab\sin C$). However, how can we find the area of a triangle if we are only given the lengths of its three sides? The answer is to use Heron's formula, which can be motivated by giving the students initially the following problems to work out. Present the following three challenges at the start of the lesson:

1. Find the area of a triangle with base and height lengths 8 and 6, respectively.

2. Find the area of a triangle whose side lengths are 3, 4, and 5.

3. Find the area of a triangle whose sides have lengths 13, 14, and 15.

For students to find the area of a triangle with base length 8 and altitude to that side has length 6, they can easily use the formula $Area = \frac{1}{2}bh$, and find the area to be $(\frac{1}{2})(8)(6) = 24$ square units. Next, for students to find the area of a triangle whose sides are lengths 3, 4 and 5, they should recognize that this is a *right triangle*, with legs 3 and 4. Thus the area is given by $(\frac{1}{2})(3)(4) = 6$ square units.

Now students are faced with the task of finding the area of a triangle whose sides are 13, 14, and 15. This is *not* a right triangle, so no side can be thought of as an altitude. There must be another way to find the area. Here is where the students should recognize a void in their knowledge of finding the area of a triangle. Again, the way it is presented in a rather dramatic fashion will aid tremendously to making this motivational technique be most effective.

One of the first student reactions is that, since 3–4–5 are sides of a right triangle, what about 13–14–15? Sadly, the triangle is *not* a right triangle. After trying to find the area (and failing to do so), the students should realize that there should be a method for finding the area of any triangle given the three sides. Hopefully you can supply it for them. This leads right into a lesson on Heron's formula for the area of a triangle. The formula was developed by the famous Greek mathematician, Heron of Alexandria (10–70 A.D.). The formula he discovered involves the use of the semi-perimeter, (s) of the triangle, which. is equal to $(\frac{1}{2})(a + b + c)$, where the three sides are lengths a, b and c. Heron's formula is $Area = \sqrt{s(s-a)(s-b)(s-c)}$.

In this the challenge presented, $a = 13$, $b = 14$, $c = 15$. Therefore, the area of the triangle[6] is $\sqrt{21(21-13)(21-14)(21-15)} = \sqrt{(21)(8)(7)(6)} = \sqrt{7056} = 84$ square units. This will now give the students a feeling of filling the void that they may have felt at the start of the lesson—one that should have motivated them to learn Heron's formula.

(2) Guide students to discover a pattern

Discovering and appreciating patterns especially with numbers is always enchanting and can serve as a motivational device. In this motivational technique, the teacher presents a contrived situation that leads students to discover a pattern. "Self-discovering" a pattern can be motivating for students because they experience positive emotions through this process. Students gain confidence when they can take ownership of their discovery of a pattern.

Examples

Topic: Introducing non-positive integer exponents

At first it may appear undaunting to students, who have just begun to understand the nature of positive-integer exponents and will probably respond to the question, "What does 5^n mean?" with a response such as "the product of n factors of 5." When asked what the nature of n is, they will probably say it is a positive integer. The teacher's response might be to question whether n has to be a positive integer. This motivator will encourage them to consider the non-positive integers: 0 and the negative integers. The teacher should now guide them through the following exercise, keeping aware that the class should be guided to appreciate the following pattern. You might show that these definitions enable an observed pattern to continue. Consider the following:

---

[6] An interesting oddity is that the length of the altitude to side 14 is 12. The numbers 12–13–14–15 are consecutive.

$$3^4 = 81$$
$$3^3 = 27$$
$$3^2 = 9$$
$$3^1 = 3.$$

Then continuing this pattern of dividing the result by 3 while decreasing the exponent by 1, we obtain:

$$3^0 = 1$$
$$3^{-1} = \frac{1}{3}$$
$$3^{-2} = \frac{1}{9}$$
$$3^{-3} = \frac{1}{27}.$$

Students should use this pattern to motivate further examination of these negative exponents. They may also look at the issue by considering $\frac{x^5}{x^5}$ (where $x \neq 0$), which equals $x^{5-5} = x^0$. Therefore, $x^0 = 1$. However, the following is a meaningless statement: "x used as a factor 0 times is 1." To be consistent with the rules of exponents, we *define* $x^0 = 1$; then it has meaning.

In a similar way, a student cannot verbally explain what x^{-4} means. What would it mean to have "x used as a factor -4 times"? Using the rules of exponents, we can establish a meaning for negative exponents. Consider

$$\frac{x^5}{x^8} = \frac{1}{x^3}.$$

By our rules of operations with exponents, we find that

$$\frac{x^5}{x^8} = x^{5-8} = x^{-3}.$$

Therefore, it would be nice for

$$x^{-3} = \frac{1}{x^3},$$

and so we can *define* it this way and our system remains consistent.
Once we arrive at

$$\frac{x^k}{x^k} = x^{k-k} = x^0,$$

we can now consider what value 0^0 should have. Using the same idea, we get

$$0^0 = 0^{k-k} = \frac{0^k}{0^k},$$

which is meaningless, since division by $0^k = 0$ is undefined. Similarly, we cannot define 0^{-k}, for this yields $\frac{1}{0^k}$, which is undefined.

Thus, students can conclude that the base cannot be 0 when the exponent is 0 or negative.

Therefore, the definitions $x^0 = 1$, $x \neq 0$, and $x^{-k} = \frac{1}{x^k}$, $x \neq 0$, become meaningful. The use of patterns motivates the students to keep exploring in this fashion to get a more solid understanding of exponents. We see here that the motivational device can be used throughout in a step-by-step fashion where it goes beyond just the beginning of the session and provides a genuine understanding of the topic.

Topic: Introduction to counting combinations

Once again, using a pattern will not only motivate the class but also take them through a genuine understanding of a concept that otherwise would not be founded in a true understanding. In that spirit, begin with the following problem for the students to ponder and then answer.

How many pairs of vertical angles are formed by 10 concurrent distinct lines?

Students will often attempt to draw a large, accurate figure, showing the 10 concurrent lines. They will then attempt to actually count the pairs of vertical angles. However, this is rather confusing, and they can easily lose track of the pairs of angles under examination. You will now have the chance to motivate them by guiding them to appreciate a pattern that evolves from this challenge. Have the students consider starting with a simpler case, perhaps using only two lines intersecting and then gradually expand the number of lines to see if a pattern emerges.

If we start with 1 line, we get 0 pairs of vertical angles.

Two lines produces two pairs of angles: marked as 1–3 and 2–4, which we showed in the first diagram of Figure 4.1.

Three lines produce 6 pairs of angles: they are 1–4; 2–5; 3–6; (1,2)–(4,5); (2,3)–(5,6); (1,6)–(3,4), which is in the second diagram of Figure 4.1.

Four lines produces 12 pairs of vertical angles: 1–5; 2–6; 3–7; 4–8; (1,2)–(5,6); (2,3)–(6,7); (3,4)–(7,8); (4,5)–(8,1); (1,2,3)–(5,6,7); (2,3,4)–(6,7,8); (3,4,5)–(7,8,1); (4,5,6)–(8,1,2), which we show in the third diagram of Figure 4.1.

Students can now be gently guided to summarize the pattern that they should have discovered as shown in the table below:

Number of Lines	1	2	3	4	5	...	N
Pairs of Vertical Angles	0	$2 = 2 \cdot 1$	$6 = 3 \cdot 2$	$12 = 4 \cdot 3$	$20 = 5 \cdot 4$	...	$n(n-1)$

By now students should realize that for 10 distinct lines, there will be $10(9) = 90$ pairs of vertical angles. This should lead into the lesson on how combinations can be counted more automatically. Students can also consider this problem from another point of view: Each pair of lines produces two pairs of vertical angles. Thus, we ask how many selections of two lines can be made from 10 lines? The answer is, of course, $_{10}C_2 = 45$. Then we get 45×2 or 90 pairs of vertical angles.

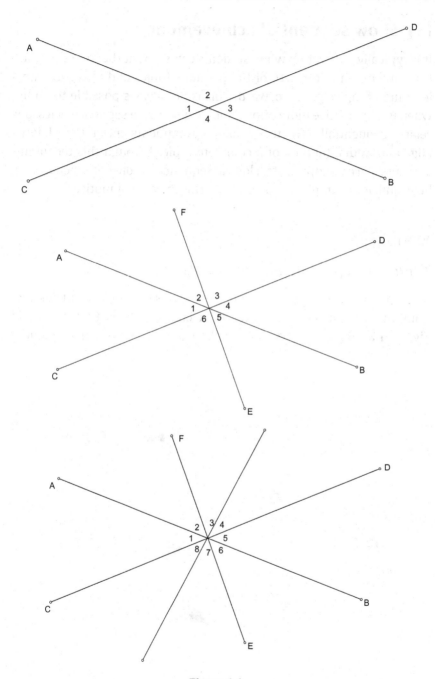

Figure 4.1

(3) Show sequential achievement

It is typically motivating when students can see that the topics they are learning are an outgrowth of the previous topics and also generate a learning of topics going forward. This is not always possible to do but when it is, it is a nice motivational tool to generate a sequence of lessons related sequentially. The technique helps students enrich their knowledge and understanding of a complete topic. Through this technique, students learn to appreciate a logical sequence of concepts. A chart may be useful, for example, when applying this method of motivation.

Example

Topic: The special quadrilaterals

In the progressive development of the various properties and descriptions of the special quadrilaterals, a chart such as that shown in Figure 4.2 will give students a feeling of progressive or sequential

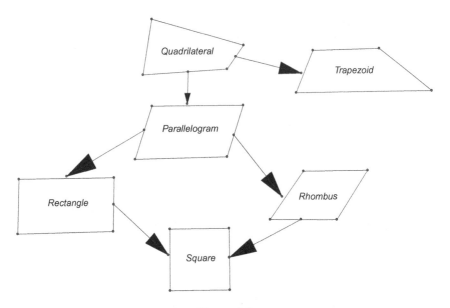

Figure 4.2

achievement, each time filling a void in their knowledge of the special quadrilaterals. This can be quite motivating as they anticipate each topic helping to complete their knowledge of the topic.

Students could be led to want to reach, sequentially, various levels of this diagrammed development. The chart should be developed carefully, with the intended purpose clearly in focus. Once again it must be stressed that the way this is used and presented is key to its effectiveness as a motivational device throughout this topic's development.

(4) Present students with a challenge

In general, challenges can be quite motivating, if done appropriately both in content and tone. When students are challenged intellectually, they often react with enthusiasm. However, great care must be taken when choosing the challenge. The challenge must lead to the lesson and the challenge must be within the reach of the students' abilities. It also helps if the challenge is short—time wise—and not overly complex. The challenge should not, for instance, be so involved that it detracts from the intended lesson. The teacher's judgment is important in this case, especially in selecting something that's appropriate for the intended student audience.

Examples

Topic: Introducing the concept of π

Prior to this lesson it would be ideal if each student could bring to the class the cardboard cylinder which is the basis for a roll of toilet paper (see Figure 4.3). Holding the cardboard cylinder up for the class to see and each student group having one to look at more closely, ask students to conjecture (without doing any measurements): which is greater, the circumference of the cylinder or the height of the cylinder? Intuitively, it is expected that students will immediately say that the height is greater.

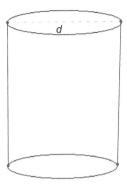

Figure 4.3

Besides using a string (or tape measure) to determine the answer to this challenge, the teacher should ask the class how else this might be determined. Remind students about the definition of π, that is, the ratio of the circumference of a circle to its diameter. This should then immediately lead them to a formula for the circumference of a circle: $C = \pi d$. Thus, they can see that the circumference is a bit more than three times the circle's diameter. Either with a ruler or simply marking off the length of the diameter and then marking off this length of diameter three times along the height of the cylinder, students will be surprised to see how much longer the circumference is that the height of the cylinder—a real surprise! This will then lead into a lesson on the nature and properties of π. Yet, the cleverness of this truly surprising result will give students a true understanding of what π is all about.

Topic: Introducing the order of operations

We offer here a challenge that will eventually lead to a solid understanding of a most important concept in mathematics. Students in any grade in middle school where the order of operations is being taught can benefit from this activity. The motivational device is to present a challenge for students to show that they can recognize the order of operations. The activity demonstrates to the class that the order of operations can yield different answers if incorrectly applied. Some of the operations students can use may not be taught until high school; thus, the activity can be revisited at any grade level.

Begin the motivational activity by providing each student or group of students with the material shown in Figure 4.4, which shows some groups of 4's that have been written in order, and the students are asked to obtain each of the natural number shown as a target using only the operation signs +, − , ×, ÷ and exponents, as well as parentheses where appropriate.

As the students work on the problems, they will notice that some students in a group may arrive at a target number in different ways. Many may not even be able to arrive at the target numbers at all. However, students will have fun with the challenge being presented. Students must decide what grouping symbols and what operational symbols can be used to arrive at the given "target number." There may be different solutions which yield the correct target number. Be sure to discuss each solution (correct or incorrect) carefully. Indicate where the incorrect order of operations was used. This activity creates the need for a lesson on the definitive *order of operations*—in effect—motivating students to get the "rules" right.

<u>Worksheet</u>

1.	4 4 4 4		Target : 2
2.	4 4 4 4		Target: 36
3.	4 4 4 4 4		Target: 16
4.	4 4 4 4 4		Target: 0
5.	4 4 4 4 4		Target: 7

Figure 4.4

Here is one set of possible answers. Note that others may also satisfy.[7]

1. $(4 \div 4) + (4 \div 4) = 2$
2. $4 \times (4 + 4) + 4 = 36$
3. $4(4 + 4) - (4 \times 4) = 16$
4. $(4 - 4) - 4(4 - 4) = 0$
5. $\sqrt{4} + \sqrt{4} + \sqrt{4} + (4 \div 4) = 7.$

For the ambitious teacher wanting to provide the class with some enrichment we offer how four 4s can be used to express the natural numbers from 1 to 20. However, it is important to remember that this is now an enrichment activity rather than a motivational one which we have already accomplished in the previous activity.

$$1 = \frac{4+4}{4+4} = \frac{\sqrt{44}}{\sqrt{44}}$$

$$2 = \frac{4 \cdot 4}{4+4} = \frac{4-4}{4} + \sqrt{4}$$

$$3 = \frac{4+4+4}{4} = \sqrt{4} + \sqrt{4} - \frac{4}{4}$$

$$4 = \frac{4-4}{4} + 4 = \frac{\sqrt{4 \cdot 4 \cdot 4}}{4}$$

$$5 = \frac{4 \cdot 4 + 4}{4}$$

$$6 = \frac{4+4}{4} + 4 = \frac{4\sqrt{4}}{4} + 4$$

$$7 = \frac{44}{4} - 4 = \sqrt{4} + 4 + \frac{4}{4}$$

$$8 = 4 \cdot 4 - 4 - 4 = \frac{4(4+4)}{4}$$

$$9 = \frac{44}{4} - \sqrt{4} = 4\sqrt{4} + \frac{4}{4}$$

---

[7] If students have learned that any number raised to the 0 power equals 1, you may provide an alternate solution such as $(4 \div 4) - (4)^{(4-4)} = 0.$

$$10 = 4 + 4 + 4 - \sqrt{4}$$

$$11 = \frac{4}{4} + \frac{4}{.4}$$

$$12 = \frac{4 \cdot 4}{\sqrt{4}} + 4 = 4 \cdot 4 - \sqrt{4} - \sqrt{4}$$

$$13 = \frac{44}{4} + \sqrt{4}$$

$$14 = 4 \cdot 4 - 4 + \sqrt{4} = 4 + 4 + 4 + \sqrt{4}$$

$$15 = \frac{44}{4} + 4 = \frac{\sqrt{4} + \sqrt{4} + \sqrt{4}}{.4}$$

$$16 = 4 \cdot 4 - 4 + 4 = \frac{4 \cdot 4 \cdot 4}{4}$$

$$17 = 4 \cdot 4 + \frac{4}{4}$$

$$18 = \frac{44}{\sqrt{4}} - 4 = 4 \cdot 4 + 4 - \sqrt{4}$$

$$19 = \frac{4 + \sqrt{4}}{.4} + 4$$

$$20 = 4 \cdot 4 + \sqrt{4} + \sqrt{4}$$

(5) Entice students with a *Gee Whiz!* amazing result

Presenting students with a "gee-whiz" moment is another motivational technique in which drama can be used with good results. Naturally, the selection of an activity of this sort must be not only appropriate for the student group considering their interest and background levels, but also needs to lead into the lesson and not distract from it. To motivate a basic belief in probability, as an early introduction into this topic, teachers can entice students with a gee whiz moment using the well-known "Birthday Problem," which challenges the students' intuition. This technique, shown in Chapter 2, page 36, should be delivered with great enthusiasm.

Examples

Topic: Introducing probability

The surprising and counterintuitive result here brings a shocking awareness to any audience. This problem, named after the French mathematician Joseph Bertrand (1822–1900), was first published in 1889, and is something that should enhance an understanding of probability.

Imagine that there are three boxes in front of you. One of these contains two gold coins, one contains two silver coins, and the third contains one coin of each type. You are invited to choose one of the three boxes at random, and then to take one of the coins from the chosen box without looking at it. When you place it on the table, you find that you have chosen a gold coin. What is the probability that the other coin in the chosen box is then also gold?

It almost seems too easy to be a question at all. There are equal numbers of silver coins and gold coins in play, so the situation must be completely symmetrical, right? So, in other words, the probability must be 50%, right? Well, no, that's not correct!

You should certainly be wary of jumping to any rash conclusions. In fact, the situation is not completely symmetric, as you have the information that the first coin you selected is gold. Seen in this light, it might even be the case that the probability of the second coin in the box, being gold as well, could conceivably be less than 50%. On the other hand, there are two boxes that contain a gold coin, so you know that the chosen box is one of these. One of these boxes has another gold coin in it, and one has a silver coin in it. So is it 50% after all?

As it turns out, the probability that the other coin in the box is also gold is actually $\frac{2}{3}$. There are several ways to see this.

Let us make it plausible that the probability of the second coin being gold is at least higher than the probability of it being silver. The easiest way to do this is by considering what happens if you play the game very often, let's say 3 million times. First of all, in each game, you choose a box. Since you are equally likely to choose any of the three boxes, you will expect to choose each box about a million times. If the chosen box was the one with two silver coins, the coin you put on the table was certainly not gold. If the box was the one with two gold coins, the coin on the table was certainly gold. Finally, if you chose the "mixed" box, the coin selected will be gold one-half of the time. That means that the coin on the table will be gold 1.5 million times, namely, all the million times you chose the gold–gold box and half-a-million times out of the ones in which you chose the mixed box. Of these 1.5 million times, the other coin is gold 1 million times, namely, all the times you chose the gold-gold box. That is 1 million times out of 1.5 million, or $\frac{2}{3}$ of the time.

Still not satisfied? Here is another way to better understand what is going on.

In Figure 4.5, we have represented the three boxes by rectangles. The gold coins are represented by the shaded circles and the silver coins by white circles. Notice that the gold–gold box contains two coins of different sizes; a large gold coin and a small gold coin.

Now, let us think of the problem conditions in a slightly different way. Instead of choosing a box first and then choosing a coin at random from within that chosen box, we can just choose any of the six coins at random. After this preliminary step, we consider the probability that the chosen coin, which we know to be gold, was in the same box as another gold coin. Of course, the probability of choosing a silver coin or a gold coin is equal if we see the setup in this way, as there are three of each in play from the beginning.

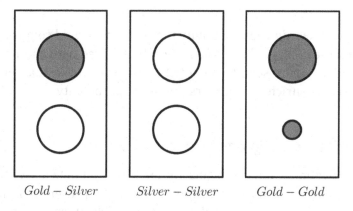

Gold − Silver   *Silver − Silver*   *Gold − Gold*

Figure 4.5

We know that we have chosen a gold coin. There are three of these in play. One of these is the one in a box together with a sliver coin. One is the large gold coin in the gold–gold box, and one is the small gold coin in the gold–gold box. Only one of these three gold coins is paired with a silver coin. The other two are each paired with a gold coin (large with small and small with large), and therefore the probability of the selected gold coin being paired in a box with another gold coin is again seen to be $\frac{2}{3}$.

As a last, quite powerful method, we can use the tools at our disposal from conditional probability. If we write P(gold|gold–gold) for the probability that we select a gold coin on the table under the assumption that the coin was taken from the gold–gold box, and we write P(gold|mixed) for the probability that we select a gold coin on the table under the assumption that the coin was taken from the mixed box, and we write P(gold|silver–silver) for the probability that we select a gold coin on the table under the assumption that the coin was taken from the silver–silver box, then Bayes' Rule, first presented by Thomas Bayes (1701–1761), gives us:

$$\frac{P(\text{gold}\,|\,\text{gold}-\text{gold})}{P(\text{gold}|\text{gold}-\text{gold})+P(\text{gold}|\text{mixed})+P(\text{gold}\,|\,\text{silver}-\text{silver})}$$

$$=\frac{1}{1+\frac{1}{2}+0}=\frac{2}{3}$$

In other words, the probability that the visible gold coin was chosen from the gold–gold box, and is therefore paired with another gold coin, is once again seen to be equal to $\frac{2}{3}$. Exposure to this thinking would surely enrich ones understanding of probability.

Topic: Introducing geometric series

This motivational device begins rather humbly but will eventually boggle students' minds. Therefore, it should be presented in a way that seems rather innocent and without any trick attached to it so that it will surprise students when they reach a result. Begin the lesson by presenting the following challenge to the class.

"Would you rather have $100,000 per day for 31 days, or

1 cent the first day

2 cents the second day

4 cents the third day

8 cents the fourth day

16 cents the fifth day

And so on for 31 days?

Experience shows that most students will opt for the first choice, namely, $100,000 for each of 31 days, since that would amount to a rather large sum of $3,100,000. The task of adding the long list of these cents amount is also something they do not care to do. At this point the teacher should ask the class how they could get the sum of these daily increasing amounts in an efficient fashion. Determining this sum easily should be sufficient motivation for students to want to learn the topic of the day: find the sum of a geometric series. As they embark on this task which of course will be the subject of the ensuing lesson, you might mention to them that they made a bad choice of money. When students eventually calculate the sum using the newly learned procedure they will find, much to their surprise, that the sum of the cents is $21,474,836.47.

Topic: Considering division by zero

When we speak of a "gee-whiz" technique to begin a lesson, the teacher might start by saying "today I'm going to prove to you that $1 = 2$." One rule in mathematics that is not emphasized enough, is that we cannot divide by zero. It is not always easy to find a situation that illustrates what happens if we mistakenly do divide by zero. The usual procedure is for the teacher to simply tell the class that they cannot divide by zero. This may not leave the class truly convinced as to why this is a forbidden division.

This motivator provides a simple situation for a class that has some basic algebraic skills. It will demonstrate what happens if we disobey this fundamental principle. You now must dramatically tell the class that you have found a proof that 1 actually equals 2. They will laugh and then you proceed to do the "proof," carefully going one step at a time, having the class supply the reason for each step.

Assume (Given)	$a = b$
(Multiply both sides of the equation by b)	$ab = b^2$
(Subtract a^2 from both sides of the equation)	$ab - a^2 = b^2 - a^2$
(Factor each side of the equation)	$a(b - a) = (b + a)(b - a)$
(Divide both sides of the equation by $(b - a)$)	$a = b + a$
(Substitute a for b on the right side of the equation)	$a = a + a = 2a$
(Divide both sides of the equation by a)	$1 = 2$

The students should experience some confusion at this point. Ask them to explain what we did that might have been wrong (if anything?). They know (obviously) that something must be wrong, since 1 cannot equal 2.

There is obviously a mathematical error. Somewhere we did something wrong. If we go back and examine each step, all appears well. So, what is the problem?

In the fifth step, we divided both sides of our equation by the quantity $(b - a)$. Since we began by assuming that $a = b$ in step 1, we are, in essence, dividing by $(a - a)$ or 0. This should show the students that strange things can happen if we divide by zero. This is a motivating way to lead to a discussion of definitions in mathematics and why they are necessary. For example, one can prove that $-1 = +1$, if we allow that $\sqrt{ab} = \sqrt{a}\sqrt{b}$, for negative a and b, which by definition is not allowed.

(6) Illustrate the usefulness of a topic (purpose)

With this motivational technique, a practical application is introduced at the beginning of the lesson. The application selected should be of genuine interest to the students (see Chapter 2 for more on the topic of getting to know your students). Again, as in the "challenge" technique, the application should be brief and not overly complicated. Remember, "usefulness" is appropriate only when students have prior knowledge of the application topic. The examples below will illustrate this point.

Examples

Topic: Introducing to the concurrency of the angle bisectors of a triangle

With a variety of students in the class it is not always easy to find examples of useful topics that would be considered interesting by all students. However, a clever teacher can present a topic which would be somewhat esoteric and yet from a style of presentation could make it appear to be useful. We will consider one such example here. Begin the class with a presentation of a problem that will motivate the relationship of the angle bisectors of a triangle—namely, that they are concurrent. The problem that the class is to grapple with is as follows: we have two wires in a field next to a lake. Each wire ends at the shoreline of the lake. Yet, if they would be extended, they would intersect in the lake. The problem is to place a third wire between these two wires so that this third wire would bisect the angle formed by the other two wires— had they been extended to the lake (see Figure 4.6) Remember, it is imperative that this be presented as a true and realistic problem with solution will lead to some interesting geometry in the ensuing lesson.

This problem will show a need to understand that the angle bisectors of a triangle are concurrent. This somewhat-practical application, which often is a challenge, will show the usefulness of the concurrency of angle bisectors in a triangle. Begin by drawing any line through the two wires and then the bisectors of the angles thus formed as shown in Figure 4.7.

Students should now realize that the desired angle bisector—the one bisecting the inaccessible angle in the lake—must contain the point, P, in which these two angle bisectors intersect.

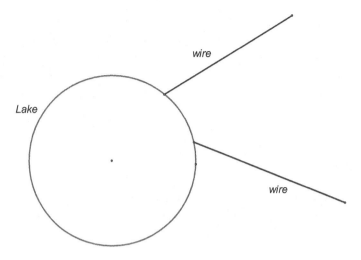

Figure 4.6

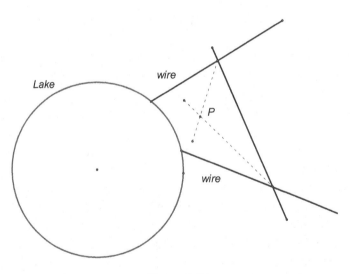

Figure 4.7

Repeating this procedure for another line that intersects the two given wires gives us the point Q (see Figure 4.8) This time the bisector of the third angle of the triangle thus formed would have to also to

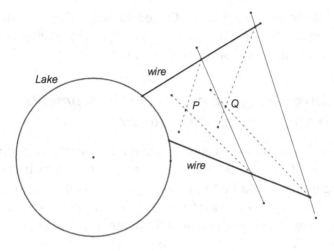

Figure 4.8

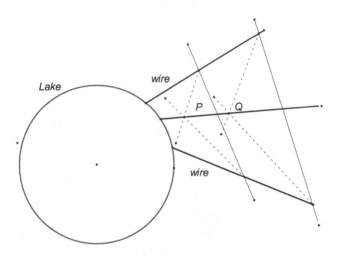

Figure 4.9

contain point Q, the point of intersection of the angle bisectors of the second triangle formed.

Students should now realize that having points P and Q, both of which lie on the bisector of the inaccessible angle in the lake will determine the desired line, shown in Figure 4.9. With this "realistic" problem students should be motivated to determine why the angle bisectors of the triangle are concurrent, especially having seen how this property improves an unusual usefulness.

Topic: Introducing the product of the segments of two intersecting chords of a circle

As the teacher enters the room to begin a lesson on the property of the product of the segments of two intersecting chords in a circle, a good beginning would be to present students with a "problem" that you had privately and needed to solve. You had a plate that broke and you needed to get a replacement of the same size. The largest piece of the broken plate was less than a semi-circle and therefore, you had the problem of determining what the diameter of the plate was. This is shown in Figure 4.10, where the dashed line represents the cracked portion.

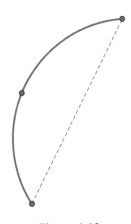

Figure 4.10

Following the lines drawn in Figure 4.11, we can measure the segments: *AE*, *BE*, and *CE*. This leads to the "cross-chords" theorem, which gives us $AE \cdot BE = CE \cdot DE$, or $6 \cdot 6 = 3x$, and $x = 12$.

Thus, students should have been motivated by having solved a practical problem using a geometric relationship that was presented in the ensuing lesson.

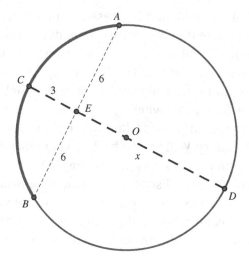

Figure 4.11

(7) Recreational motivation (puzzles, games, paradoxes)

This is a rather difficult form of motivation, largely because it has to be carefully selected to lead into the ensuing lesson and at the same time cannot be so much "fun" that that the lesson for which it is intended to motivate is overshadowed by the initial activity of the day. Using puzzles, games, and paradoxes in the classroom can be fun for students—these are recreational techniques that can be used to motivate students. Recreational activities should be brief and relatively simple since students should be able to enjoy (and master) without much effort.

Examples

Topic: Understanding percents

When the teacher presents this motivational activity, it must be done in somewhat of a story fashion as a conundrum faced by someone that students can relate to. It should be presented in a fashion that would make it appear recreational for the intended student audience. Applying it to some local concern would make it even more recreational. It also has to be presented in this "real-life-situation" tone. We will demonstrate it in a general way merely as a model.

Suppose you had a job where you received a 10% raise. Soon thereafter because business was falling off, the boss was forced to give you a 10% cut in salary. Will you be back to your starting salary? The answer is a resounding (and very surprising) NO!

This little story is quite disconcerting, since one would expect that the same percent increase and decrease should take you where you started from. This is intuitive thinking, but wrong! Convince yourself of this by choosing a specific amount of money and trying to follow the instructions.

Begin with $100. Calculate a 10% increase on the $100 to get $110. Now take a 10% decrease of this $110 to get $99 – $1 less than the beginning amount.

Students may wonder whether the result would have been different if we first calculated the 10% decrease and then the 10% increase. Using the same $100 basis, we first calculate a 10% decrease to get $90. Then the 10% increase yields $99, the same as before. So, order apparently makes no difference. This will lead the teacher into a discussion of percentages with a now-motivated class. By the way, there are number of clever techniques the teacher can use during the lesson to cover a number of situations related to the above problem.[8]

Topic: Applications of algebra counterintuitive peculiarities

This example has to be presented in such a way that students will be led to something apparently inexplicable but will get clarity through

---

[8] For a more detailed discussion of one such technique, see A.S. Posamentier, *Math Charmers: Tantalizing Tidbits for the Mind.* Guilford, CT: Prometheus Books, 2003, pp. 96–98.

simple algebra. When we teach a lesson on algebraic applications, we often leave students questioning the real need for these applications. To counteract this negative attitude, begin this class with the following recreational example. This lovely little problem will show your students how some clever reasoning along with *algebraic knowledge* of the most elementary kind will help them solve a seemingly "impossibly difficult" problem.

Here is the problem that might be worked on by students individually or in groups, whichever fits the class more appropriately.

You are seated at a table in a dark room. On the table there are 12 pennies, 5 of which are heads up and 7 are tails up. (You know where the coins are, so you can move or flip any coin, but because it is dark you will not know if the coin you are touching was originally heads up or tails up.) You are to separate the coins into two piles (possibly flipping some of them) so that when the lights are turned on there will be an equal number of heads in each pile.

Your first reaction is "you must be kidding! How can anyone do this task without seeing which coins are heads or tails up?" This is where a most clever (yet incredibly simple) use of algebra will be the key to the solution.

After students have had ample time to get a bit frustrated, the teacher may lead the discussion through pointed questioning. Have them begin by separating the coins into two piles, of five and seven coins each. Then flip over the coins in the smaller pile. Now both piles will have the same number of heads! That's all! Your students will think this is magic. How did this happen? Well, this is where algebra helps understand what was actually done.

When you separate the coins in the dark room, h heads will end up in the 7-coin pile. Then the other pile, the 5-coin pile, will have $5 - h$ heads. To get the number of tails in the 5-coin pile, we subtract the number of heads ($5 - h$) from the total number of coins in the pile, 5, to get: $5 - (5 - h) = h$ tails (see Figure 4.12).

5-coin Pile	7-coin Pile
5 – h heads	h heads
5 – (5 – h) tails	
= h tails	

Figure 4.12

The piles after flipping the coins in the smaller pile	
5-coin Pile	**7-coin Pile**
5 – h tails	h heads
h heads	

Figure 4.13

When you flip all the coins in the smaller pile (the 5-coin pile), the (5 – h) heads become tails and the h tails become heads. Now each pile contains h heads! (see Figure 4.13).

This absolutely-surprising result will show your students how the simplest algebra can explain a very complicated reasoning exercise and hopefully motivate the students to place greater value on algebraic applications.

(8) Tell a pertinent story

A carefully selected and presented story can be a powerful motivational technique—in practically any setting. For teachers who are familiar with the story they plan to tell their students in a classroom setting, however, it is critical to not rush through the presentation. Of course, a bit of drama helps when using the story-telling technique. Remember that a carefully prepared method of presentation of the story is almost as important as the content of the story itself. The most appropriate selection of stories will also rely upon the teachers' knowledge of his or her students.

Examples

Topic: Finding the sum of an arithmetic series

Begin this lesson in a storytelling fashion, where the teacher should take time to properly dramatize the situation about to be shared with the class. While a student in elementary school, in the 18th Century the young Carl Friedrich Gauss (1777–1855), who later went on to become one of the greatest mathematicians in history, had as his teacher, Mr. Buettner, who one day wanted to keep his class occupied. To do so, he simply asked the class to use their slate boards (apparently, they didn't use paper for classroom exercises) and find the sum of the first 100 natural numbers. The students did what was asked of them. Namely, they began to sum the numbers $1 + 2 + 3 + 4 + 5 + 6 + \cdots$ until they reached 100. One student did not do the assignment this way and he finished immediately. That was young Carl Gauss. He decided to approach the problem in a different fashion. Rather than to add the numbers improper order:

$1 + 2 + 3 + 4 + 5 + 6 + \cdots + 98 + 99 + 100$, he decided to add the numbers in pairs:

$1 + 100 = 101$

$2 + 99 = 101$

$3 + 98 = 101$

$4 + 97 = 101$

And so on, until he reached

$48 + 53 = 101$

$49 + 52 = 101$

$50 + 51 = 101$

Of course, he did not write all this, having realized that each pair had a sum of 101, and that there were 50 such pairs. Therefore, he simply multiplied $50 \times 101 = 5,050$ to get the sum of the numbers that Mr. Buettner requested. When young Carl raised his hand to offer the answer the teacher told him to keep quiet and do the work as was requested, anticipating that would take about ½ hour for the students to finish the task. At the end of the allotted time, Carl turned out to be the only student in the class to get the right answer.

An effective teacher would now need to use this technique to develop the lesson. However, when some teachers use this story as a lead into the lesson on finding the sum of an arithmetic series, they tend to revert to the typical textbook method for developing the formula for finding such a sum, thus defeating the fine motivational aspect of this delightful little story—one that Gauss was proud to repeat in his older days.

To use this motivational story to develop a formula for the sum of an arithmetic series one might begin by representing the series in general terms as:

$$a + (a + d) + (a + 2d) + (a + 3d) + \cdots + (a + (n - 3)d) + (a + (n - 2)d)$$
$$+ (a + (n - 1)d)$$

By adding the first and the last terms and then the second and the next-to-last terms and so on, we get:

$$a + (a + (n - 1)d) = 2a + (n - 1)d$$
$$(a + d) + (a + (n - 2)d) = 2a + (n - 1)d$$
$$(a + 2d) + (a + (n - 3)d) = 2a + (n - 1)d$$

It becomes clear that there is a pattern developing—each pair yields the same sum. This is the same pattern that Gauss got when he added the numbers in pairs (as shown above). When we seek the sum of n numbers in this arithmetic series beginning with the first number a and with a common difference between terms of d, we have to sum $\frac{n}{2}$ pairs. Therefore, the sum of the series is: $\frac{n}{2}(2a + (n - 1)d)$.[9]

We stress here that if one uses a story to motivate a lesson, then the essence of the story should not be lost by then using another technique to develop the concept to be taught. This unfortunately

---

[9] This formula is often seen as $S = \frac{n}{2}(a + l)$, where $l = a + (n-1)d$.

happens if a teacher tells this story and then reverts to the textbook's development—if it is different from that used here—to derive the formula for the sum of an arithmetic series. In short, using the story to motivate a lesson should end up being a part of the lesson itself.

Topic: Introduction to the Pythagorean theorem

There are times when the most unexpected things can be entertaining such as the following. The teacher might begin this lesson on the Pythagorean theorem by asking the students what do the following men have in common Pythagoras, Euclid and U.S. President James A. Garfield? The answer is that they each proved the Pythagorean theorem in a different way. It should be noted that two of the most famous presidents of the United States were fond of mathematics. Washington was adept at surveying and spoke favorably about mathematics and Lincoln was known to carry a copy of Euclid's Elements in his saddle bag while he was still a lawyer. This, too, can enhance the storytelling to lead into the discussion of the Pythagorean theorem.

You might then tell the story of Garfield's proof and then show it. Garfield discovered the proof about 5 years before he became president. During this time, 1876, he was a member of Congress and hit upon the idea during a conversation about mathematics with other members of Congress. The proof was later published in the *New England Journal of Education*.

After this story, with other embellishments that the teacher wishes to use here, the Garfield proof can be presented as follows.

To begin **President James A. Garfield's Proof** we consider right $\triangle ABC$ with $m\angle C = 90$, as shown in Figure 4.14. We let $AC = b$, $BC = a$, and $AB = c$. We need to show that $a^2 + b^2 = c^2$.

Select D on BC so that $BD = AC$ and CBD Consider $DE \perp CBD$ so that $DE = BC$. We can show that quadrilateral $ACDE$ is a trapezoid. Also, Area $\triangle ABC$ = Area $\triangle BED$, and $AB = BE$.

Area trapezoid $ACDE = \frac{1}{2}\ CD(AC + DE) = \frac{1}{2}\ (a + b)(a + b) = \frac{1}{2}\ (a + b)^2$

Area $\triangle ABE = \frac{1}{2}\ AB\ BE = \frac{1}{2}\ c^2$. Also, Area $\triangle ABC = \frac{1}{2}\ AC\ BC = \frac{1}{2}\ ab$.

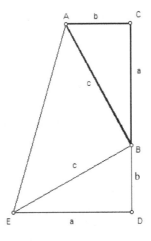

Figure 4.14

However, Area trapezoid $ACDE$ = Area $\triangle ABE$ + 2Area $\triangle ABC$. Substituting, we get

$$\frac{1}{2}(a+b)^2 = \frac{1}{2}c^2 + 2\left(\frac{1}{2}ab\right)$$

$$(a+b)^2 = c^2 + 2ab$$

and it follows that $a^2 + b^2 = c^2$. So, we have shown that President James A. Garfield was also able to prove the Pythagorean theorem in a rather simple and clever fashion.

(9) Actively engage students in justifying mathematical curiosities

This motivational technique attempts to actively engage students in justifying a mathematical curiosity. The students should be familiar with the mathematical curiosity before the teacher challenges them to justify it. While this may take more time than normally allotted for a motivational activity, to proceed before students understand the concept is counterproductive.

Examples

Topic: Application of digit problems in algebra

It is important to make a presentation of the mathematical curiosity dramatized by the enthusiasm that a teacher should express in its delivery. In that spirit, begin the class in the unusual way of asking every student to select a three-digit number, where the unit and hundreds digit are not the same. This mind-boggling activity shows number properties that are truly exceptional, leaving students with a strong desire to understand why this is so—hence a highly motivating the students.

We will provide you with a simulated version of what is to be done with the class. In bold print are the instructions for the students to follow, while our model number in boxed.

Choose any three-digit number (where the unit and hundreds digit are not the same).

> We will do it with you here by arbitrarily selecting: **634**

Reverse the digits of this number you have selected.

> We will continue here by reversing the digits of 634 to get: **436**

Subtract the two numbers (naturally, the larger minus the smaller)

> Our calculated difference is: **634 − 436 = 198**

Once again, reverse the digits of this difference.

> Reversing the digits of 297 we get the number: **891**

Now, add your last two numbers.

> We then add the last two numbers to get: 198 + 891 = **1089**

Each student's result should be the same as ours, even though each student's starting number was different. Students will probably be astonished with the result that regardless of which number was selected at the beginning, the same result was reached by all, namely, 1089. How does this happen? Is this a "freak property" of this number? Did we do something illegitimate in our calculations?

Can they assume that any number we chose would lead us to 1089? How can we be sure? Well, they could try all possible three-digit numbers to see if the number 1089 will appear at the end. That would be tedious and not particularly elegant. This is where the teacher can embark on the lesson—applications of algebra with digit problems.

Following is a possible method to better understand what was going on during our calculation and which can serve as a guide for the teacher. We shall represent the arbitrarily selected three-digit number, **htu** as $100h + 10t + u$, where h represents the hundreds digit, t represents the tens digit, and u represents the units digit. Let $h > u$, which would be the case either in the number you selected or the reverse of it. In the subtraction, $u - h < 0$; therefore, take 1 from the tens place (of the minuend) making the units place $10 + u$. Since the tens digits of the two numbers to be subtracted are equal, and 1 was taken from the tens digit of the minuend, then the value of this digit is $10(t - 1)$ The hundreds digit of the minuend is $h - 1$, because 1 was taken away to enable subtraction in the tens place, making the value of the tens digit $10(t - 1) + 100 = 10(t + 9)$.

We can now do the first subtraction:

$$
\begin{array}{lll}
100(h-1) & +100(t+9) & +(u+10) \\
100u & +10t & +h \\
\hline
100(h-u-1) & +10(9) & +u-h+10
\end{array}
$$

Reversing the digits of this difference gives us:

$$100(u - h + 10) + 10(9) + (h - u - 1)$$

Now adding these last two expressions gives us:

$$100(9) + 10(18) + (10 - 1) = \underline{1089}$$

So, now students can see how algebra enables us to inspect the arithmetic process, regardless of the number. This is something that should be strongly emphasized by a teacher to motivate students to appreciate algebra and see it as a process beyond a mechanical task required in a mathematics curriculum.

Teachers should take every opportunity to enrich instruction where possible. One such opportunity presents itself with the particular beauty in the number **1089**. As a little extra, you may want to entertain the class with another oddity of this now famous number, 1089.

Let's look at the first ten multiples of 1089.

$$1089 \cdot 1 = 1089$$
$$1089 \cdot 2 = 2178$$
$$1089 \cdot 3 = 3267$$
$$1089 \cdot 4 = 4356$$
$$1089 \cdot 5 = 5445$$
$$1089 \cdot 6 = 6534$$
$$1089 \cdot 7 = 7623$$
$$1089 \cdot 8 = 8712$$
$$1089 \cdot 9 = 9801$$

Ask students if they notice a pattern among the products? They should look at the first and ninth products. They are reverses of one another. The second and the eighth are also reverses of one another. So, the pattern continues, until the fifth product is the reverse of itself, known as a palindromic number.

(10) Use teacher-made or commercially-prepared materials

In this final technique, motivation can be achieved by presenting the students with concrete material of an unusual nature. These materials can be custom made by the teacher or a they can be commercially made products. There are a number of options, in various price-points, available online. Materials must be selected and reviewed thoughtfully—and their use in a lesson carefully planned—so that the materials motivate students for the lesson as opposed to detracting attention from the lesson.

Examples

Topic: Concept of similar triangles

When a teacher brings an unusual instrument into the mathematics classroom in and of itself it presents a bit of awe and curiosity among students. This can serve as a motivational device. The pantograph shown in Figure 4.15 is a linkage instrument that is used to draw

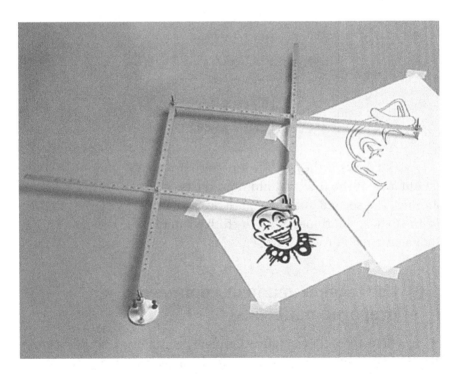

Figure 4.15

Source: www.rockler.com/product.

similar figures. It can be obtained in toy stores or drawing stores as it has been sold to children to allow them to draw (or trace) cartoon characters in an enlarged form. It can also be constructed with strips of cardboard and fasteners.

The pantograph consists of four bars hinged at four points, with one point fixed and another point having a pencil. Holes are provided on the bars to allow for size adjustment. Tracing triangles and noting the ratio of similitude by the adjustments on the bars will be a good lead in to the study of similarity. Time should be taken to explain how the instrument works and how it leads into and supports similarity.

Topic: Introducing the concept of a function

Once again, the motivation here is the unusualness of the beginning of this lesson. The teacher brings to the class a toy bow and arrow set just to dramatize the motivational activity presented here. Just the sight of the unusual items will arouse—or motivate—students to discover what the ensuing lesson is going to be about. The following is actually a strong introduction to the concept of functions that has been motivated by the bow and arrow set and will give students a genuine understanding of the concept of functions by the story that follows:

In mathematics, finding concrete analogues to represent abstract concepts is not always easy. One example where a physical model can be used to explain an abstract concept is in the development of the notion of a function.

We will use the model of a bow shooting arrows at a target. The arrows will represent the *domain* and the target represents the *range*. The bow (and its aiming) is the *function*. Since an arrow can only be used once[10] we know that the elements in the domain can be used only once. The bow can hit the same point on the target more than once. Therefore, points in the range can be used more than once.

---

[10] Actually, a gun and bullets would be a better analogue than the bow and arrow, since there the bullet can *really* be used only once. For this illustration, make it clear that the arrow, once shot, cannot be used again.

This is the definition of a function: a *mapping* of all elements of one set onto another, with the elements of the first set used exactly once. Some points on the target may never be hit by an arrow, yet all the arrows must be used. Analogously, some elements in the range may not be used, but all elements in the domain must be used. Or conversely, through a mapping (or a "pairing") of all elements in the domain, some elements in the range may not be used.

When all points on the target (the range) are hit[11] then the function (or mapping) is called an *onto function.*

When each point on the target is used only once, then the function is called a *one-to-one function.*

When each point on the target is used exactly once (i.e. once and only once) then the function is called a *one-to-one onto function*, or may be called a *one-to-one correspondence.*

Using the bow-shooting-arrows-to-a-target analogy to represent the concept of a function enables the learner to conceptualize this abstract notion in a way that should instill permanent understanding of the concept of a function.

In summary, remembering a few general rules for using these motivational techniques will make them more effective:

- Motivation should be brief but applied daily.
- Motivation should not be overemphasized; it is a means to an end, not an end in itself.
- Motivation should elicit the aim of the lesson of the class; this is an excellent way of determining the effectiveness of the technique.
- Motivation should be appropriate to the grade level and interests of the students.
- Motivation should draw on the motives actually present in the students.

---

[11] Obviously, in reality, an infinite number of arrows would be required, so it must be appropriately simulated.

Chapter 5

Know How to Engage Your Students

"How can I most effectively engage my students?" This is the perennial question asked by educators all across the globe. What techniques draw students into discussions or entice them to engage with the material? What exciting new technology or ideas will mesmerize and engage the students, resulting in a fascination, for example, with quadratic equations? Despite advances in technology, including a wide variety of options and methods for using technology in the classroom and beyond, and despite a plethora of pedagogical theories about engagement, teachers continue to struggle with drawing students out and engaging them in the lesson material. Part of the problem with putting too much stock in technology is that technology is simply ancillary—or at the very best—a means to engagement. We have learned that technology is not the one grand solution to solving the problems of short attention spans, boredom, and lack of imagination or interest. In fact, some researchers claim that technology can be the cause of these issues.

It is always helpful to have a toolbox of pedagogical approaches to engagement, ready to put into play when necessary. All the years spent studying education theories and practicing methods in student-teaching has certainly prepared you to put some of those best practices to the test, so to speak. This chapter on knowing how to engage your students offers another tool for engagement that has not been adopted by the field of education. It is a low-tech and inexpensive, yet innovative, technique that actually had its start in the field of aviation—in the United States Army Air Corps. The technique explored in this chapter is the "checklist," which is an easy way to double-check a process or a piece of work that helps avoid mistakes and improve outcomes.

In 1935, at Wright Field in Ohio, the United States Army Air Corps held a tryout among aircraft companies for its new bomber. The aircraft manufacturing company Boeing entered its B-17 in the tryout. The B-17 plane was a complicated one and even though the pilot was highly trained and experienced, after the plane took off, it stalled, crashed, and burst into flames. This was all due to a simple routine step that had been forgotten. Because of this tragedy, pilots began to adopt the use of checklists.[1]

The new Boeing B-17 plane was *complicated*, just as education is complicated. In the 21st century, we live in a fast-moving global world. Our world today is even more complicated than the world of 1935—technologically, socially, environmentally, economically, medically, and politically. The classroom is a microcosm of the larger world with a myriad of complications. Once in the classroom, we are confronted with the worlds—the personal experiences and environments—of each of our students. Educators are expected to set learning standards for their class and to help students achieve those standards. Educators are expected to present material relevant to grade level, and to, hopefully, put this in context of students' lived realities. The classroom has become very complicated indeed! Exploring the ways that the checklist can be adapted to an educational environment could prove instructive for engaging students in the classroom as well as for the more complicated administrative tasks that we must undertake with our colleagues and supervisors.

---

[1] A. Gawande, *The Checklist Manifesto: How to Get Things Right.* New York, NY: Picador, 2011.

Since adopting the checklist system in 1935, the field of aviation has seen incredible results. In part, because of the checklist, when we travel by air there is now only a one in 29.4 million chance of being killed on any particular commercial flight.[2] Pilots routinely perform a visual survey as they walk around the aircraft with a checklist. Once inside the cockpit, if it is a commercial flight, they review a checklist with their co-pilot. It is not widely known that Captain Chesley "Sully" Sullenberger, the pilot who landed a commercial jet on the Hudson River after a flock of geese took out both engines, credits the adherence to a checklist in his ability to bring down the aircraft safely, saving all the passengers on board. It was his decades of experience and extensive training that taught him about the importance of the checklist. The dramatic life-or-death situation Captain Sullenberger and his co-pilot found themselves in was extremely *complicated*. The checklist provided a clear path forward.

Not surprisingly, the construction industry has adopted the checklist as well. Engineers, architects, and construction managers employ various checklists throughout the construction process, especially when the project has many different moving parts—that is, when the construction project is highly complicated. Just as in the field of aviation, any missteps or simple oversights in construction can be catastrophic. Consulting checklists means that the construction industry has a relatively low building-failure rate: one in 50,000 structures partially or fully collapse.[3] Despite years of training and extensive experience even the most intelligent people make mistakes. Humans have unreliable memories, we are flawed decision makers, and our attention sometimes wanders at crucial moments. The checklist process provides a cognitive net, a system to ensure high quality work.

While we do not have to be as concerned as pilots or engineers about catastrophic events that have the potential to cause death, as educators we deal with extremely complicated educational systems and the larger bureaucratic bodies that inform these systems. We must

---

[2] *Ibid.*
[3] *Ibid.*

understand and adhere to complicated and ever-changing state and federal regulations concerning the achievement of educational goals, outcomes, and assessment. In addition, we play a role in complicated systems of communication within our school districts in which students can sometimes fall through the cracks. Within the classroom environment we are also confronted with complicated material since, in almost every field, knowledge has exploded bringing with that exponential growth and a focus on specialization. In order to deal with this expansion of knowledge and the complexities that result from that expansion, we must learn to collaborate and communicate more effectively with our colleagues.

The practice of the checklist provides us with an opportunity to teach our own students about collaboration. Through an attempt to teach collaboration to our students, we can engage them with a tool not commonly used in education, namely, the checklist. We can use the checklist on both a macro and micro level: on a broad scale across administrative leaders, staff, and teachers, and in the classroom to engage our students with the material and with each other. From the examples already mentioned, we see that the use of the checklist in aviation and engineering has proven extremely effective. There is no doubt that it has saved lives. Another field has more recently adopted the checklist with great success, and that is the field of medicine.

When the Harvard surgeon Dr. Atul Gawande was asked by the World Health Organization to work on a project to better understand the high rate of complications from surgery and the relatively high death rate, he turned to other fields to research how they dealt with highly complex processes. Through this research, he discovered the value of the checklist in both aviation and engineering as well as in something as low-tech as cattle herding (yes, cowboys use a checklist!). In medicine, it appeared that despite the high level of intelligence, training, and experience; the intense focus on specialization; and new advances in technology, surgeons still made careless mistakes. Those mistakes kept complications from surgery and death rates elevated. One of the examples Gawande cited was a failure to have a sufficient amount of antibiotics in the surgical theatre to administer to the patient (antibiotics cut the rate of infection by half when used in a timely fashion). Items on the checklist that Gawande and his team reviewed before surgery included a check for antibiotics supply, to verify the patient's name, the type of operation to be performed, whether there was an extra blood supply available in case of an emergency, and a basic introduction of the surgical team, some of whom might not have ever worked together before the surgery, among other checklist items.

What is quite astonishing are the results of using this checklist process before surgery. In the seven hospitals around the world that implemented the use of the checklist, Gawande and his team found that the rate of major complications from surgery dropped by 35%. The drop in the number of deaths from surgical procedures dropped by an incredible 47%. Considering these outcomes, it is not surprising that this surgical checklist process is now standard operating procedure in most United States hospitals. These results were generated through an attempt to understand and address a particular problem in medicine through the innovative use of a technique that originated from the aviation and engineering fields. One of the key points to take away from this project is that a low-tech, simple process such as the checklist can have a significant impact. Any field that has complicated systems or processes could most likely benefit from the use of a checklist designed specifically to address problems that arise out of human error or inattention or unreliable memory. By the way, Gawande worked with the lead safety engineer at the Boeing aircraft manufacturing company to help devise his checklist for the surgical theatre.

Using Checklists in an Educational Environment

Clearly, checklists work well in fields that have complicated processes such as engineering, aviation, medicine, and even cattle-herding. But how can we use the checklist in education? What would a checklist used in a classroom (in a lesson plan or to manage other processes) or even for an administrative purpose look like? For a straightforward, practical checklist, first, identify important projects or critical decisions that need to be made in your classroom or school. Second, focus on identifying these five components:

1. The checklist should have a clear "pause point." This is a moment when it is clear that the teacher or student needs to pause and consult the checklist.

2. The checklist should be speedy. It should take less than 60 seconds to complete, and the list should have 5–9 "critical" items.

3. The checklist should be a supplement to existing knowledge and expertise.

4. The checklist should be field-tested and continually updated (past failures and lessons learned).

5. Maintain records for assessment purposes.

If, for instance, you are introducing a complex mathematical problem to your students, perhaps a problem that has many steps, try creating a checklist for that problem. Mathematics problems are actually a perfect fit with the checklist process since if one step is overlooked or done incorrectly in a mathematics problem, it is impossible to come to the right solution. The pause moment—that moment in which your students pause and consult the checklist—could be precisely at the point where many students get lost or confused in the problem. As a teacher, you know what that sticking point is. It is that place where the problem becomes particularly complicated. If you have your checklist already created for this problem, put the list on the board/screen and start to go through the list with your students. Engage them by asking them to identify whether those 5–9 "critical" points have been completed. If these points have not been addressed in their work, it will result in either poor quality work or have a negative impact on their work.

Students can work with a partner or in small groups as they go through the checklist. Or, together as a class, you can review the critical points one by one. This should generate engagement from students who already have a familiarity with the mathematical concept, since you will have already presented the concept to the class. However, even though students will have a familiarity with the material, they may still stumble when figuring out the complex problem on their own. In this situation, the checklist is a great way to help students to practice, both in the classroom with other students or individually. In addition, if you are assigning homework, the students can refer to the checklist for a reminder of those critical points they need to check off to get a correct answer to the problem.

For example, a checklist can be very helpful in doing a formal deductive proof in geometry. Some of the key points that might be on the checklist would be as follows:

1. What is being asked to be proved?

2. What are some of the key points that are directly given?

3. Am I using all the given information?

4. What facts can be gotten directly from the given information? (For example, if an isosceles triangle is shown in the figure, a pair of equal base angles would evolve directly.)

5. What techniques have I encountered that would be useful to use in this proof? (This could include identifying parallel lines and their properties, or identifying congruent triangles, or properties of special triangles such as isosceles or equilateral triangles, etc.)

6. Working backwards, what are the characteristics of the end product to be proved? (For example, if one is trying to prove a quadrilateral is a parallelogram, what minimum information is needed to establish a parallelogram, such as both pairs of opposites sides parallel, or both pairs of outside equal in length, or one paragraph sides equal and parallel, etc.)

7. Have I reached a conclusion based on logical deduction?

There may be other questions that one might want to include in this checklist based on teacher recommendation and personal experience.

Another example, where a checklist could be quite useful, is in a mathematical problem that is somewhat tricky to solve. We provide one such here, merely as a rather extreme example. Problem solving experiences are not necessarily done merely to solve the problem at hand. Rather, the challenge of solving an unusual problem can open up a person's thinking processes for future analogous encounters. This of course can be streamlined nicely with a checklist. The problem is merely a vehicle for presenting a technique for solution. It is from the types of solutions that one really learns problem solving, since one of the most useful techniques in approaching a problem to be solved is to ask yourself checklist items such as the following:

1. Have I ever encountered such a problem before?

2. Have I duly noted all the given information?

3. Working backwards, what am I expected to conclude? (Speed, time, distance, etc.)

4. Which one of these is dependent on the others and how?

5. Focusing on the end result, how can I avoid being distracted by any redundant information given?

Don't be deterred by the relatively lengthy reading required to get through the problem. You will be delighted (and entertained) with the unexpected simplicity of the solution.

Two trains, serving the Chicago to New York route, a distance of 800 miles, start towards each other at the same time (along the same tracks). One train is traveling uniformly at 60 miles per hour and the other at 40 miles per hour. At the same time a bumblebee begins to fly from the front of one of the trains, at a speed of 80 miles per hour towards the oncoming train. After touching the front of this second train, the bumblebee reverses direction and flies towards the first train (still at the same speed of 80 miles per hour). The bumblebee continues this back and forth flying until the two trains collide, crushing the bumblebee. How many miles did the bumblebee fly before its demise?

It is natural to be drawn to find the individual distances that the bumblebee traveled. An immediate reaction is to set up an equation based on the famous relationship (from high school mathematics): "rate × time = distance." However, this back and forth path is rather difficult to determine, requiring considerable calculation. Just the notion of having to do this can cause serious frustration. Even if you were able to determine each of the parts of the bumblebee's flight, it is still very difficult to solve the problem in this way.

A much more elegant approach would be to solve a simpler analogous problem (one might also say we are looking at the problem from a different point of view). We now begin to employ the checklist. It should be noted that each person will probably create a checklist based on their own experience. So here we can see how the step-by-step procedure to an elegant solution can be reached and can be applied to one's personal checklist.

We seek to find the *distance* the bumblebee traveled. If we knew the *time* the bumblebee traveled, we could determine the bumblebee's distance because we already know the *speed* of the bumblebee. Having two parts of the equation "rate × time = distance" will provide the third part. So, having the *time* and the *speed* will yield the distance traveled, albeit in various directions.

The time the bumblebee traveled can be easily calculated, since it traveled the entire time the two trains were traveling towards each other (until they collided). To determine the time, t, the trains traveled, we set up an equation as follows: The distance of the first train is $60t$ and the distance of the second train is $40t$. The total distance the two trains traveled is 800 miles. Therefore, $60t + 40t = 800$, so $t = 8$ hours, which is also the time the bumblebee traveled. We can now find the distance the bumblebee traveled, using the relationship, rate × time = distance, which gives us $(8)(80) = 640$ miles.

It is important to stress how to avoid falling into the trap of always trying to do what the problem calls for directly. At times a more circuitous method is much more efficient. Lots can be learned from this solution. You see, dramatic solutions are often more useful than traditional solutions, since they provide an opportunity "to think out of the box." Here, we can see how the checklist, in a sense, guides us from falling into a trap by having us approach the problem logically, in this case in somewhat of a reverse path.

Be creative in thinking about other beneficial ways that the checklist might be used in your classroom to encourage engagement with and among students. Is there a point in the school day when students tend to drift off, or have a difficult time focusing? The period just after lunch (or recess?) might be one of those times for many students. Think of the checklist as a way to "check-in" with students and to re-engage them after a break. This could be that mid-day "pause" moment in which they go through a checklist to determine if they are ready to proceed to the next lesson, for instance. In a sense, they could be that pilot doing a visual survey of the jet, checking to make sure that the aircraft is in good condition and that all systems are go. Depending on the grade level, this checklist could include critical points such as: Do you have a pencil and paper ready, or perhaps your laptop, etc.? If you plan on using a PowerPoint or merely a whiteboard projection, is the technology in place to avoid any time delay? What do you need for your next period lesson?

If you need to prepare a checklist for end-of-day activities the list would be different. Those critical points might include questions (or reminders) about preparing to go home: Did you remember to put your homework exercises in your bookbag? Have you straightened your desktop area? Have you fed and watered the classroom plants? Do you have the _____ form to give to your students' parents? As time goes on, you will find it necessary to update checklists. In fact, it is absolutely critical to continually refine and revise your checklists based on past failures and lessons learned. Different classes might require a variety of other checkpoints, depending upon the room location, the class level and the assignments involved. This is both an analytical and creative process, however, in which you might well include your students.

Including students in the creation of checklists is essential to student engagement and can be fun. You can first engage students with the checklist concept by creating them yourself and helping them to understand the value in the checklist process. Designing and constructing visually pleasing checklists is one way to pull students into the process. For instance, design a checklist that includes a large box next to each critical point. As you discuss and complete the points, call individual students or pairs of students up to draw a big "X" in the box with a colorful marker. They could even draw a bold line through the entire critical point. Devise something exciting and appropriate for the student's grade-level and ask them to put their own creative touch on the checklist when deciding how they want to mark the point as having been completed. The only materials needed for an exercise like this are paper and large markers. The possibilities to be explored in these types of checklist projects are endless and are only limited by your imagination.

In addition to creating visually appealing checklists—and encouraging your students to participate in this process—students can also create the content of the checklists. Depending on the grade-level of your students, this exercise should encourage the kind of analytical thinking that allows them to fully assess the problem at hand: Why is this a pause point? What should the critical points be for this particular case? You should only ask this of your students after you have already created checklists and used them with your students in class. They should have a firm understanding of the concept of the "pause point," for example. When you reach a complicated point in a lesson, you can then pause and ask your students whether this might be a good place to incorporate the checklist process. What is it about this particular point that necessitates a checklist or would benefit from using one? Encourage the students to describe how a checklist would or would not benefit those who need to figure out the problem or who need to be reminded of the important steps.

Engaging students through the use of checklists and their participation in the development of checklists could turn into a useful practice that they take with them to the next grade level or, hopefully, incorporate in some other ways into their life. When teachers take time to develop an exercise with their students—whether it is a checklist process or a creative project—there is a greater potential for engagement. When students are able to take an ownership position, there is a greater potential for engagement. When students can successfully complete complex mathematical problems, their confidence is bolstered and there is a greater potential for engagement.

The checklist is a tool that we can use to model problem-solving and analytical thinking for our students as well. The checklist process allows students to identify pause points (or trouble areas) and then to develop a list of critical points. When these critical points are addressed in the checklist process, students have a greater chance of producing good quality work. This process can also be helpful with complicated problems that need to be broken down into smaller sections so that students can successfully tackle the problem one section at a time.

Here is a very simple problem with an even simpler solution. Yet the solution most people will come up with is much more complicated. Why? Because they look at the problem in the psychologically traditional way: the way it is presented and played out. Here is where using a checklist could be quite helpful. Try the problem yourself (don't look below at the solution) and see whether you fall into the "majority-solvers" group. In setting up a checklist:

1. You would want to list what you are given.

2. You would want to consider an alternative solution, such as perhaps, working backwards.

3. You would want to focus on what you are seeking, and perhaps avoiding distractions.

The problem: *A single elimination (one loss and the team is eliminated) basketball tournament has 25 teams competing. How many games must be played until there is a single tournament champion?*

Typically, the majority-solvers will begin to simulate the tournament, by taking two groups of 12 teams playing the first round, and thereby eliminating 12 teams (12 games have now been played). The remaining 13 teams play, say 6 against another 6, leaving 7 teams in the tournament (18 games have been played now). In the next round, of the 7 remaining teams, 3 can be eliminated (21 games have so far been played). The four remaining teams play, leaving 2 teams for the championship game (23 games have now been played). This championship game is the 24th game. Now we revert to the checklist method.

A much simple way to solve this problem, one that most people do not naturally come up with as a first attempt, is to focus only on the losers and not on the winners as we have done above. We ask the key question (point 3): "How many losers must there be in the tournament with 25 teams in order for there to be one winner"? The answer is simple: 24 losers. How many games must be played to get 24 losers? Naturally, 24. So there you have the answer, very simply done. Now most people will ask themselves, "Why didn't I think of that?" The answer is, it was contrary to the type of training and experience we have had. Becoming aware of the strategy of looking at the problem from a different point of view may sometimes reap nice benefits, as was the case here. This is where a checklist might help a student in solving a problem.

A lengthy reading problem can put some people off, for fear that they won't even understand what is being asked. This is where a checklist could prove useful. Once past the statement of the problem, it is very easy to understand, but quite difficult to solve by conventional means.

The elegant solution which could be supported by a checklist method offered later—as unexpected as it is—almost makes the problem trivial. However, our conventional thinking patterns will likely cause a confusing haze over the problem. Don't despair. Give it a genuine try. Then read the solution provided here. We begin by stating the problem.

We have two one-gallon bottles. One contains a quart of red wine and the other, a quart of white wine. We take a tablespoonful of red wine and pour it into the white wine. Then we take a tablespoon of this new mixture (white wine and red wine) and pour it into the bottle of red wine. Is there more red wine in the white wine bottle, or more white wine in the red wine bottle?

To solve the problem, we can figure this out in any of the usual ways—often referred to in the high school context as "mixture problems"—or we can use some clever logical reasoning and look at the problem's solution using a short checklist as follows:

1. With the first "transport" of wine there is only red wine on the tablespoon.

2. On the second "transport" of wine, there is as much white wine on the spoon as there is red wine in the "white-wine bottle."

3. This may require students to think a bit, but most should "get it" soon.

The simplest solution to understand and the one that demonstrates a very powerful strategy is that of *using extremes*. This is something that ought to be included in a checklist when solving some problems. We use this kind of reasoning in everyday life when we resort to the option: "such-and-such would occur in a worst case scenario, so we can decide to . . . ."

Let us now employ this strategy with our checklist for the above problem.

1. To do this, we will consider the tablespoonful quantity to be a bit larger.

2. Clearly the outcome of this problem is independent of the quantity transported.

3. So, we will use an *extremely* large quantity.

4. We will let this quantity actually be the *entire* one quart.

That is, following the instructions given the problem statement, we will take the entire amount (one quart of red wine), and pour it into the white-wine bottle. This mixture is now 50% white wine and 50% red wine. We then pour one quart of this mixture back into the red-wine bottle. The mixture is now the same in both bottles. Therefore, there is as much white wine in the red wine bottle as there is red wine in the white wine bottle!

We can consider another form of an extreme case, where the spoon doing the wine transporting has a zero quantity. In this case the conclusion follows immediately: There is as much red wine in the white-wine bottle as there is white wine in the red-wine bottle, that is, zero!

Carefully presented, this solution can be very significant in the way students approach future mathematics problems and even how they may analyze their everyday decision-making.

Ultimately, the checklist process encourages a proactive approach to identifying all kinds of problems and then offers the steps needed for solving those problems.

Checklists for Administrative Processes

The checklist could prove useful in clarifying and improving various administrative processes among teachers and administrators as well. In our modern and highly complicated school systems, clarity in communication is essential. Supporting institutional processes and procedures in place that ensure the well-being and safety of all students, both academically and behaviorally, is also critical. Yet, we know that sometimes, despite our extensive training and experience, we make bad decisions because of an unreliable memory or lack of attention to detail. In our complicated educational systems, we can see how some students might slip through the cracks or somehow just get lost in the system. For instance, you might want to consider the checklist for both academic and behavioral student cases at your school. Ask yourself, is there some identifiable pause point in certain administrative processes? Where are the mistakes happening concerning student academic and behavioral support? Is it the referral system for counseling? Or perhaps you can identify a pause point when students move from one grade level to the next. A concise, well-constructed and relevant checklist would vary from school to school. It should be custom designed to address the specific problems that teachers and administrators are attempting to solve.

The checklist also has another benefit: It has the potential to increase collaboration between teachers and administrators and staff. This increased collaboration can improve systems of communication, which is at the heart of any top-performing school (or classroom). In addition to encouraging collaboration and communication, the checklist enables us to see that despite all the knowledge that exists, if we do not apply that knowledge properly, we fail. In an increasingly complicated world, intelligent, experienced, and well-trained people can make simple mistakes. That is the story of the complicated B-17 Boeing jet that despite the pilot's training and experience crashed in the try-out. Or the story of the surgeon with years of training and experience who mistakenly operates on the wrong side of the patient. In both cases, the mistakes could have been avoided with the use of a simple, low-tech and well-designed checklist.

In the field of education, we are not dealing with the extreme life and death situations described in the aviation, engineering, and medicine cases. However, as teachers we are tasked with helping students understand complicated subjects, our students are complicated, and the school systems in which we teach seem to be getting more complicated each year. As we think about the tools available to us when trying to engage our students, we might want to be aware of those pause points where we could implement the checklist process. The checklist is not appropriate for all situations, of course, but using it to assist students as they grapple with complicated problems can be useful in a number of ways. It is an opportunity to teach students how to identify a pause point—a place in the process where we need to stop and pay close attention to the next steps. Once they understand how the checklist process works, students can create their own checklists, which is a relevant exercise in analytical and creative thinking. A mathematical problem may not be a life or death situation, but we are preparing students to be critical thinkers who participate in society and who, perhaps, go on to train as surgeons, pilots, or engineers.

Chapter 6

Empower Students to Become Leaders

Businesses have learned that word of mouth advertising, customer referrals, and good reviews are much more powerful than promotional messages that come directly from the company's public relations office or an advertising firm. Now, more than ever, business marketing strategy relies on social media, which has become a core instrument for increasing brand awareness and exposure through direct engagement with consumers. As more and more people search for product information and make purchases online, the influence of social media is amplified exponentially. This chapter explores the various ways we can create an environment in which it is possible for students to become leaders—much like social media "influencers"—through a variety of methods used in social media marketing and adapted for the classroom.

Many people enjoy sharing their opinions and post their interesting finds online. They also like to see—and are influenced by—content posted by their peers. Research indicates that 71% of consumers who have had a positive experience with a brand on social media are likely to recommend the brand to their friends and family.[1] About 49% of

---

[1] A. Arnold, 4 Ways Social Media Influences Millennials' Purchasing Decisions. *Forbes*, December 22, 2017.

consumers depend on influencers' recommendations on social media.[2] The most effective advertising is word-of-mouth reviews and recommendations from people not affiliated with their company. Word-of-mouth advertising has no cost to the business, yet it is invaluable. With the use of social media, this now happens online through consumers' engagement and the creation of content. Some people have actually become well-known for their reviews, recommendations, and "expertise" with certain products. These people are known as "influencers," and businesses seek them out and pay quite a bit of money to access influencers' platforms, which can consist of hundreds of thousands or even millions of followers. To the businesses that hire influencers, the influencers' followers represent potential customers.

Unlike traditional advertising methods such as print or television, social media has the potential to engage viewers by encouraging them to proactively create content. As far as the company is concerned, the more interaction, the better. Businesses want to be able to engage and interact with consumers outside of their official website, customer support options, and regular communication channels. When a company posts content and an influencer reposts or responds to the content, it reaches everyone in the influencer's social media universe—people who may have not even been interested in the brand. This unofficial, viewer-propelled advertising contributes to what is known as the "multiplier effect." The term comes from the field of economics and refers to injections (investments, government spending, exports) into the economy that cause a new demand for goods and services. This demand stimulates further rounds of spending and can lead to a bigger eventual final effect on output and employment. In other words, one person's spending is another person's income. In marketing, this concept can be understood as the output of social media content that is amplified by the response to and/or the sharing of that content by viewers. The original content is multiplied usually through an influencer or, occasionally, when a post goes "viral."

---

[2] Digital Marketing Institute, Twenty Influencer Marketing Stats That Will Surprise You. Retrieved on May 5, 2020 at: DigitalMarketingInstitute.com.

How do we understand this concept of the multiplier effect in the classroom? And, how can we use the multiplier effect to empower students to become leaders, or in the language of social media, "influencers," for learning? Teaching leadership or providing opportunities for students to lead helps them with personal and social skills that are in demand and will help them navigate all aspects of life. Using social media platforms to develop those leadership skills utilize an environment and language with which your students are probably already familiar.[3] Using the environment and language of social media provides an opportunity to discuss a number of topics related to leadership characteristics as well as leadership in practice: community consciousness, accountability, responsibility, positivity, accepting criticism, courage, perseverance and dedication, decision-making skills, collaboration, negotiation, and risk-taking, among others.

Since most of your students are already online and may be using social networking technologies, your classroom can benefit from a few of the best practices used in social media marketing. These best practices have the potential to prompt some of the students who do not usually emerge as leaders to become influencers in the class. To put it in the language of social media marketing: Your classroom is your *network*. Within that network you have *first-level* connections (those students who you have identified as potential leaders/ influencers). Your *reach* is your *second-level* connections, the rest of the students in the classroom. In this second-level, however, there will be students who can be cultivated as influencers—and these are the students who may ultimately have even more influence on the rest of the class. In marketing, for example, if 10% of your network interacts with your content, then that content becomes exposed to their network. If 5% of their network interacts with your content, it becomes exposed to their network and so on down the line. With each extension, the multiplier for your content goes well beyond three (or even six) degrees of separation.

---

[3] This approach follows what has been covered in previous chapters about meeting your students in their world—teaching to their interests.

Supposing you would like to have your students popularize the field of probability before you enter it in your formal classroom instruction. You might engage them to participate in an experiment which is often referred to as the famous "birthday problem", and which we already presented in Chapter 1.

While the birthday problem showed us that some probability results are quite counterintuitive, here we will show a very controversial issue in probability that also challenges our intuition. There is a rather famous problem in the field of probability that is typically not mentioned in school, yet it has been very strongly popularized in newspapers, magazines and even has at least one book[4] entirely devoted to the subject. It is one of these counterintuitive examples that gives the deeper meaning to understanding the concept of probability.

This example stems from a long-running television game show "Let's Make a Deal" that featured a rather curious problematic situation. As part of the game show a randomly-selected audience member would come on stage and be presented with three doors. She was asked to select one, with the hope of selecting the door which had a car behind it, and not one of the other two doors, each of which had a donkey behind it. There was, however, an extra feature in this selection process. After the contestant made her initial selection, the host, Monty Hall, exposed one of the two donkeys, which was behind a not-selected door—leaving two doors still unopened. The audience participant was asked if she wanted to stay with her original selection (not yet revealed) or switch to the other unopened door. At this point, to heighten the suspense, the rest of the audience would shout out "stay" or "switch" with seemingly equal frequency. The question is what to do? Does it make a difference? If so, which is the better strategy (i.e. the greater probability of winning) to use here? Intuitively, most would say that doesn't make any difference, since there are two doors still unopened, one of which conceals a car and the other a donkey. Therefore, many folks would assume there is a 50–50 chance that door which the contestant initially selected is just as likely to have car behind it as the other non-exposed door.

---

[4] Jason Rosenhouse, "The Monty Hall Problem: The Remarkable Story of Math's Most Contentious Brain Teaser", New York: Oxford University Press, 2009.

Let us look at this entire situation as a step-by-step process then correct response should become clear. There are *two donkeys* and *one car* behind these three doors. The contestant must try to get the car. Let's assume that she selects Door #3. Using simple probability thinking, we know the probability that the car is behind door number 3 is $\frac{1}{3}$. Therefore, probability that the car is behind either door number 1 or door number 2 is then $\frac{2}{3}$. This is important to remember, as we move along.

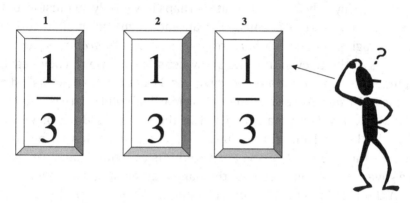

Knowing where the car is hidden, the host, Monty Hall, then opens one of the two doors that the contestant did *not* select and exposes a donkey. Keep in mind the probability that the car is behind one of these two doors is $\frac{2}{3}$.

He then asks the contestant: "Do you still want your first-choice door, or do you want to switch to the other closed door"? Remember, the combined probability of the car being behind door number 1 and door number 2 is $\frac{2}{3}$. Now with door number two exposed as not having the car behind it, the probability that the car is behind door number 1 is still $\frac{2}{3}$, while we recall the probability that the car is behind door number 3 is still $\frac{1}{3}$. Therefore, the logical decision for the contestant is to switch to door number 1.

This problem has caused many an argument in academic circles and was also a topic of discussion in the *New York Times*, and other popular publications as well. John Tierney wrote in *The New York Times* (Sunday, July 21, 1991) that "perhaps it was only an illusion, but for a moment here it seemed that an end might be in sight to the debate raging among mathematicians, readers of *Parade* magazine, and fans of the television game show Let's Make a Deal. They began arguing last September after Marilyn vos Savant (1946–) published a puzzle in *Parade*. As readers of her 'Ask Marilyn' column are reminded each week, Ms. vos Savant is listed in the Guinness Book of World Records Hall of Fame for 'Highest I.Q.,' but that credential did not impress the public when she answered this question from a reader." She gave the right answer, but still many mathematicians argued.

Although this is a very entertaining and popular problem, it is extremely important to understand the message herewith imparted and one that should by all means have been a part of the school curriculum, not only to make probability more understandable, but also more enjoyable.

Before we get into examples of how the multiplier effect might unfold in a classroom environment, it is worth repeating that students are not consumers and we are not advocating the marketing of material to our students as is the practice in business. The education of our students involves much more than the transactional relationship between the seller and the buyer in the world of commerce. As educators, we strive to support the development of the whole student. There are some lessons to be learned, however, from the field of social media marketing, and we might consider how to transform some of these best practices in a way that is both relevant and appropriate in education.

The potential for adopting these best practices became apparent in during the COVID-19 pandemic in the spring of 2020. In March, when COVID-19 swept through New York City, the City University of New York (CUNY) made the decision to transfer all in-person courses to online learning—as did most colleges and universities across the nation. While faculty had used various learning platforms as a supplement to teaching in some in-person courses over the years, many had never gone completely online. What some faculty discovered in the move to a 100% online teaching environment was surprising.

By mid-March, for instance, many faculty members had spent almost 2 months meeting in-person twice a week with 34 students in, for example, an introductory business course. By then, they had identified those students who were leaders or "influencers" in the class. As usual in a classroom setting, there emerge a small group of highly engaged students who were willing and able to communicate their ideas and who were consistently prepared for a lively class discussion. Once classes went online, some of those students who were willing to engage in an in-person classroom setting, actually receded into the background and a new group of students emerged. This new group was engaged with the material and regularly posted on discussion forums. Their "voices" were prominent in a way that they had not been in the in-person classroom. It appeared that the medium of online learning enabled this new group to engage in a way they had not engaged in the in-class setting. The online platform seemed to be the preferred communication medium for this group of students and allowed them to contribute in a way that had not previously.

The COVID-19 pandemic forced a situation in which, during one semester, faculty could compare the engagement and performance of students from 100% in-class to 100% online, which was an unusual opportunity, to say the least. While strictly anecdotal, this experience illustrates how a different group of students from the same class emerged as leaders—or at least much more engaged in two different classroom settings. Faculty started to think about those students in all of their in-person classes who did not present themselves as leaders but, if given the right environment, they could emerge as leaders/influencers. If there are potentially two groups of leaders in some classes, how might faculty take advantage of this? How might this new information, if used properly, change the dynamic among the entire class?

The characteristics of an effective leader and social media influencer overlap, and both the leader and the influencer can take advantage of the multiplier effect. While you identify and call on leaders in your classroom, you may choose to use the language with your students that is often used in social media marketing when discussing the concept of the multiplier effect and leader/influencer. Take this opportunity to enter their world of social networking to engage them while setting up guidelines and boundaries regarding civility and community-building online. Help them develop leadership skills by illustrating those connections between a strong leader and a well-known influencer with a large following. A leader must communicate effectively, consider other people's opinions, manage their own emotional response, and make smart decisions. The use of social media networking in your classroom can help students develop these skills. You will recognize the leaders in your classroom faster than expected. Moreover, you can encourage all students in the class to work on the skills that will turn them into future leaders—or influencers. Through the use of social media networking platforms, you work toward achieving leadership goals related to:

- **Communication skills:** Cooperation, problem-solving and listening skills, deal with conflict situations.
- **Social awareness skills:** Consider other peoples' opinions and show empathy and interest.
- **Emotional management skills:** Manage stress and motivate one-self to work towards particular goals.
- **Self-awareness skills:** Identify and express feelings through the process of collaboration.
- **Decision-making skills:** Make reasonable choices that everyone agrees with in group. (Best to allow people to take ownership of decisions made).

In your own classroom, you will be able to identify these "naturally-occurring" classroom leaders. They are those eager students whose arm shoots up before you even finish asking the question. While they seem to consistently do well with verbal communication, this may not be the case with their written communication or with analytical problem-solving. You might be tempted to focus exclusively on these students as your first-level connections. However, you would be missing the opportunity to take advantage of, and develop the leadership and communication skills of, another group of students in your class. This group is the second-level connections who might do extremely well communicating through online platforms. The first-level connections in your class already influence other students. The challenge is to use the second-level connections to influence even more students in the class. Not only will you be able to create a large group of influencers, but you will also demonstrate to the students that there is more than one type of leader. In our classroom example in which we described the insight from switching from in-person to online settings, the first-level connections are with those students who actively participated through verbal engagement. The second-level student connections may have a more understated style of communication even though they have mastered the material, and they are not as outspoken in an in-person class as the first-level students.

Using the multiplier effect concept from social media marketing to encourage and promote student leadership could be implemented in a variety of ways and will be explored in detail in the rest of this chapter. However, before you start exploring how you might use social media platforms such as Facebook, YouTube, or Instagram in your class, you need to make sure that you fully understand any guidelines that may have been instituted by the school district regarding students' privacy and social media. There are a number of ways that you can also ensure privacy through the platform itself. Once the school district's policy regarding social media use is understood, you can take advantage of the privacy settings or special uses of social media to protect your students' (and your own) identity. If you decide to use Facebook for the social networking platform, for example, choose a Private Facebook Group.

There are two types of Private Facebook Groups to consider for your class: Visible Facebook Groups (formerly known as Closed) and Hidden Facebook Groups (formerly known as Secret). Both groups will provide varying levels of control over who joins, who can participate, and who sees the content. A Visible Facebook Group, while "private," is searchable. That means that the name of the group and the names of its members are visible and can come up in an online search. However, the content posted in this group is not accessible to the public—only the current members are able to view group content. On the other hand, a Hidden Facebook Group—as its name implies—is hidden completely from search; only members can find the group online. In a Hidden Facebook Group, the group name, its members, and content are completely hidden from non-members. A school's policy on social media will dictate which Facebook Group to use, or whether one will be able to use Facebook at all. If you do use Facebook Group, when sending the information about joining the Facebook Group home to parents/guardians, make sure to include a permission form that will allow images and videos of the student to be posted. Even if it is a Private Facebook Group (Visible or Hidden), some parents may prefer that their child's image not be posted on social media out of safety or privacy concerns. This permission form will give them an opportunity to opt out.

If you do use Facebook Group (or some other network-type platform such as a blog), there are a number of ways you can encourage student engagement through your two groups of student influencers. Set some ground rules for the students on how the class will use the platform. As the teacher, you are the administrator of your class' social networking group and provide rules regarding the type of material that is appropriate for posting and what is acceptable and not acceptable in language and tone when responding to a post. This is a wonderful opportunity to discuss online community-building, civility, and cyber-bullying. Considering the amount of time children and young adults spend online, practicing responsible online citizenship is invaluable.

In terms of course material, you can use the platform to help rein-force lessons taught during class. For instance, if you have a mathe-matic lesson plan that covered a complicated problem—try posting a video in which you demonstrate the various steps at the board or create a lesson on Prezi or PowerPoint to post. Your students would be able to access this video later at home when they are doing their homework, and it can serve as a visual reminder of how to solve the problem.

Identify some students from both groups of class influencers and ask them to demonstrate how to solve certain mathematical problems while being videotaped (to post on the class Facebook Group or blog later) or in-person during class, when appropriate. Supporting their classmates in learning encourages students to develop leadership skills such as community consciousness, responsibility, and critical and creative problem-solving. In addition, and to the point of the concept of the multiplier effect, also identify students who are not the obvious influencers in the class, but who have shown through work that they have turned in, that they understand the material and could be cultivated to be influencers. As mentioned earlier in this chapter, the first-level connections in your class are the students who are the naturally-occurring leaders, and this is obvious because of their in-class engagement. These students have leadership—and influencer—potential. The second-level connections are the rest of the class, and in this group you will find students who consistently get high grades but perhaps they do not participate that much in class. In addition to cultivating your first-level connection influencers, focus on drawing out those influencers in the second-level connections group through involvement in your online educational platforms— this becomes the place where they shine and exert their influence. In addition, the content that gets posted on the class' Facebook Group or blog will be open to comments and "likes," which provides a forum for peer-to-peer interaction, support, community building, and influencing other students.

Try grouping your first- and second-level influencers with their classmates in teams and in one-on-one sessions. Connect the assignment you give to the Facebook Group platform or blog and require students to upload their completed assignment to the class site. Part of any assignment can be to ask students to evaluate their experiences through written or recorded reviews, like those seen on Yelp, for instance. Students can work in teams to create content for the class Facebook Group, such as a review, a short video, a solution to a problem, among many other topics. One of the benefits of using social media in the classroom is providing students with the opportunity to connect their work socially and academically to a wider world (in this case, the "safe" world of the classroom). When students learn to collaborate and critique each other's assignments, they develop leadership skills that help in practically every aspect of their life. During these activities, student influencers from the first- and second-level connection emerge and the multiplier effect ripples through the "network" of the classroom. Students are already familiar with social media "influencers"—those people whom others want to emulate. By discussing influencers and leadership development in class, you can set a high standard for influencer characteristics. Ask students who their favorite influencer is on social media or in any media. Discuss leadership qualities and how those might align or clash with their online role models.

Encourage students to become influencers in their own classroom. Setting aside time in class throughout the week for students to engage in activities related to the Facebook Group or classroom blog provides opportunities for influencers to explain a concept to other students, for example. Social media platforms have a broader application that goes beyond posting information. You can use social media networking platforms to highlight stellar student work, and as the administrator you can encourage and moderate student responses. Give your students the responsibility of designing a class competition that uses the platform to post mathematical challenges or games. Also try creating a place on the platform where student influencers can offer guidance or tutoring to their peers—this can cover academic subjects as well as social issues. An advice column might be a creative way to nurture community-building and civility.

There are problems that look deceptively simple, but are not. Here is one that has baffled entire high school mathematics departments and can serve as an interesting discussion point for students as they investigate a plausible solution on the classroom online platform. This problem certainly makes for an interesting investigation and provides good interaction among students. Once we expose the solution, some students are disappointed in not having seen the solution right from the start. So here it is. Have students work with the problem without looking at Figure 6.2, as it will give away the solution.

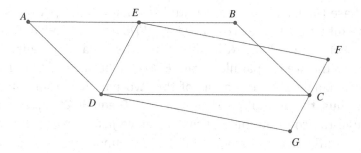

Figure 6.1

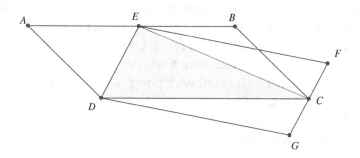

Figure 6.2

In Figure 6.1, point *E* lies on *AB* and point *C* lies on *FG*.

The area of parallelogram *ABCD* = 20 square units. Find the area of parallelogram *EFGD*.

(*Note*: Typically, the area of a parallelogram is found by taking the product of its height and base lengths. This is not possible here so another procedure will have to be found.)

Students should be allowed to ponder the solution and drive various approaches before exposing the following solution.

Begin by drawing *EC* as in Figure 6.2.

Since triangle *EDC* and parallelogram *ABCD* share a common base *DC* and a common altitude (a perpendicular from point *E* to *DC*), the area of triangle *EDC* is equal to one-half the area of parallelogram *ABCD*. Similarly, since triangle *EDC* and parallelogram *EFGD* share the same base (*ED*), and the same altitude to that base (a perpendicular from point *C* to *ED*), the area of triangle *EDC* equals one-half the area of parallelogram *EFG*D. Now, that since the area of parallelogram *ABCD* and the area of parallelogram *EFGD* are both equal to twice the area of triangle *EDC*, the areas of the two parallelograms must be equal. Thus, the area of parallelogram *EFGD* equals 20 square units. Although the solution method that we have just shown is not often used, it is effective and efficient. But more importantly this problem will generate a great deal of discussion among students and then in amazement at the simplicity of the solution.

Furthermore, this problem can also be solved quite elegantly by solving a simpler analogous problem (without loss of generality). Recall that the original given conditions were that the two parallelograms had to have a common vertex (*D*), and a vertex of one had to be on the side of the other as shown with points *E* and *C*. Now, let us suppose that point *C* coincided with point *G*, and point *E* coincided with point *A*. This satisfies the given condition of the original problem and makes the two parallelograms coincide. Thus, we can once again conclude that the area of parallelogram *EFGD* = 20 square units.

Students could also look at this last solution as one of using extremes. That is, they might consider point *E* on *AB*, yet placed at an extreme, such as on point *A*. Similarly, we could place point *C* on point *G* and satisfy all the conditions of the original problem. Here again we have the problem made trivial, in that the two parallelograms overlap. You might have students look back and recall how difficult they perceived the problem at the start.

We offer here another problem that will provide ample discussion among students as they attempt a solution. When the problem is presented, the symmetry makes it look disarmingly simple but just wait.

Find the numerical value of the following expression:

$$\left(1-\frac{1}{4}\right)\left(1-\frac{1}{9}\right)\left(1-\frac{1}{16}\right)\left(1-\frac{1}{25}\right)\cdots\left(1-\frac{1}{225}\right)$$

The usual "knee-jerk reaction" to the question is to get each of the terms "simplified." That is, doing the indicated subtraction within each pair of parentheses (e.g. $1 - \frac{1}{4} = \frac{3}{4}$) to get the following:

$$\left(\frac{3}{4}\right)\left(\frac{8}{9}\right)\left(\frac{15}{16}\right)\left(\frac{24}{25}\right)\cdots\left(\frac{224}{225}\right)$$

It is not unusual that students may choose to change each fraction to a decimal (with a calculator!) and multiply the results (again with a calculator). It is obviously a very cumbersome calculation and such a solution would certainly not be considered elegant.

An alternative and far more elegant method would be to organize the data in a different way—recognizing that the terms are comprised of perfect squares. This will permit students to look at the problem from a different point of view, with the hope of getting some sort of pattern that will enable them to simplify their work.

$$\left(1^2 - \frac{1}{2^2}\right)\left(1^2 - \frac{1}{3^2}\right)\left(1^2 - \frac{1}{4^2}\right)\left(1^2 - \frac{1}{5^2}\right)\cdots\left(1^2 - \frac{1}{15^2}\right)$$

Students may then notice that they can factor each parenthetical expression as the difference of two perfect squares,[5] which yields:

$$\left(1 - \frac{1}{2}\right)\left(1 + \frac{1}{2}\right)\left(1 - \frac{1}{3}\right)\left(1 + \frac{1}{3}\right)\left(1 - \frac{1}{4}\right)\left(1 + \frac{1}{4}\right)\left(1 - \frac{1}{5}\right)\left(1 + \frac{1}{5}\right)\cdots$$
$$\left(1 - \frac{1}{15}\right)\left(1 + \frac{1}{15}\right)$$

when they then do the subtraction within each pair of parentheses they will get.

$$= \left(\frac{1}{2}\right)\left(\frac{3}{2}\right)\left(\frac{2}{3}\right)\left(\frac{4}{3}\right)\left(\frac{3}{4}\right)\left(\frac{5}{4}\right)\left(\frac{4}{5}\right)\left(\frac{6}{5}\right)\cdots\left(\frac{13}{14}\right)\left(\frac{15}{14}\right)\left(\frac{14}{15}\right)\left(\frac{16}{15}\right)$$

A pattern should now become evident, and they may see the advantage multiplying pairs of fractions as indicated.

$$\left(\frac{1}{2}\right) \ \left(\frac{3}{2}\right)\left(\frac{2}{3}\right) \ \left(\frac{4}{3}\right)\left(\frac{3}{4}\right) \ \left(\frac{5}{4}\right)\left(\frac{4}{5}\right) \ \left(\frac{6}{5}\right)\left(\frac{5}{6}\right) \cdots \left(\frac{14}{13}\right)\left(\frac{13}{14}\right)$$
$$\left(\frac{15}{14}\right)\left(\frac{14}{15}\right) \ \left(\frac{16}{15}\right)$$

As a result, surprisingly, we are left with $\left(\frac{1}{2}\right)\left(\frac{16}{15}\right) = \frac{8}{15}$. Here students will be able to appreciate the power of a well-organized set of data.

The ways in which you might use social media in your class are limited only by your imagination. In addition to sharing material with your students and their parents, you can make announcements, send reminders, humorous stories, and allow for engagement between students in a discussion forum dedicated to a particular topic. Students can also be tasked with selecting relevant content for classmates to reference and share. These ideas all have the potential to increase communication between teachers and parents as well as increase the feel of community in the classroom with your students. Much of this is fairly easy to do with access to a smartphone or even with conference call apps installed on a laptop in the classroom. Mobile devices can display HD-quality content, faster and more efficiently than ever before. In fact, mobile Internet usage surpassed desktop back in 2016 and that trend continues.

Social media, of course, are free—but, unfortunately, not free of advertising messages. There may be other educational online platforms that you have access to that are more appropriate for your students. If your school district does not allow the use of social media as part of class projects, they may purchase (or be willing to provide) an alternative—there are many options currently on the market. In addition, there are several free online products, which have a network option built into the product. Remember that a blog could also be used for these social networking activities, and could replace expensive platforms and even the free ones such as Facebook and YouTube.

Worldwide, in 2020, more than 4.5 billion people use the Internet, and 3.8 billion people use social media. About 60% of the world's population[6] (heavily concentrated in wealthy counties) are online and the numbers increase every year,[7] and 96% of students with Internet access report using social networking technologies.[8] Once your students join the workforce, they will be expected to use social networking technologies—Google Chat, Microsoft Team, or Zoom, for example.[9] The strategies covered in this chapter will help a greater number of students develop skills that enable responsible and civil use of social media while also building leadership skills through peer-to-peer interaction and social networking interaction.

---

[6] Approximately 7.5 billion.

[7] *Digital 2020: Global Digital Overview*. Accessed 5/1/2020 at: https://wearesocial.com/blog/2020/01/digital-2020-3-8-billion-people-use-social-media.

[8] Infographic, Best Masters in Education.

[9] By the time this book goes to press, these technologies will most likely have developed into more sophisticated versions. If students understand the concept behind social networking, they will be able to keep up with new technologies.

Chapter 7

Keeping Up with Teaching Technologies

Slates, chalkboards, paper, and pencils. When we think of technology, these are probably not the first things that come to mind. In the 21st century, most people might even consider these classroom classics outdated, if not obsolete. Yet in the mid-19th-century in the United States, the chalkboard was a new teaching technology in the classroom[1] and the pencil followed close behind, once the widespread production of affordable paper became available for use in the classroom in the mid-1800s. An understanding of technology[2] depends entirely on the period being discussed—new technology in the mid-1800s was a chalkboard that enabled the teacher to demonstrate mathematics or language problems to an entire classroom of students. In the United States in the early 1800s, the traditional classroom, which involved full-class instruction by subject-matter specialist

---

[1] S. Krause, "Among the Greatest Benefactors of Mankind": What the Success of Chalkboards Tell Us about the Future of Computers in Classrooms. *The Journal of the Midwest Modern Language Association*, 33(2), 6–16 (2006).

[2] The American sociologist Read Bain's definition of technology is widely accepted: "Technology includes all tools, machines, utensils, weapons, instruments, housing, clothing, communicating and transporting devices, and the skills by which we produce and use them." Bain, Read, Technology and State Government. *American Sociological Review*, 2(6), 860–874 (1937).

teachers, was influenced by the British educator Joseph Lancaster (1778–1838). Lancaster developed a system that included the chalkboard, a model for classroom lay-out, the organization of material, and whole class instruction.[3] Fast forward to 2020, and new technology might look like sophisticated educational software incorporating augmented reality (AR) glasses that could potentially give students the experience of talking to Isaac Newton or Albert Einstein while sitting at their desk in your classroom.

One way we identify and understand technology is through its application—the application of tools, knowledge, and materials to solve problems and to extend human capabilities. Technology is not one thing that exists on its own; the essence of technology is in its application. The chalkboard enabled teaching in front of large groups of students by providing the teacher with a way to demonstrate how to find the solution to mathematical equations, for example. Using that technology enabled all the students to see and follow along. It was the application of the chalkboard that made it a relevant technology in the mid-1800s. What looks like sophisticated technology now, however, will overtime look as rudimentary as the chalkboard or paper and pencil seem to us today.

The history of the development of technology in education is a fascinating trek through time, and a close examination of this history reveals that technology has always been at the forefront of human education. In 1870, for example, the new technology of the "magic lantern" was all the rage. The magic lantern projected images (paintings, prints, photographs) on glass plates and was commonly used for entertainment purposes. It is actually an early version of the slide projector, which came much later. Educators picked up on the potential of using the magic lantern in the classroom, and by 1918, the Chicago public school system had approximately 8,000 slides circulating through its classrooms.[4] A few years later, in 1920, innovative

---

[3] S. Krause, "Among the Greatest Benefactors of Mankind": What the Success of Chalkboards Tell Us about the Future of Computers in Classrooms. *The Journal of the Midwest Modern Language Association*, 33(2), 6–16 (2000).

[4] The Learning Machines. *The New York Times*, September 15, 2010. A History of Classroom Technology: The Primitive Classroom. Purdue Education Blog, Purdue University. Retrieved on June 01, 2020 at: https://online.purdue.edu/blog/education/evolution-technology-classroom.

educators start using the new technology of radio to provide on-air classes to anyone within listening range. This application of radio is actually quite similar to the creation and application of Massive Open Online Courses (MOOCs), first introduced in 2008 and widely available by 2012.[5] In their own time, radio and MOOCs were both new technologies that attempted to reach and educate a wider audience. They both attempted to solve a similar problem and extend human capabilities, yet both were developed almost a century apart.

The ball-point pen, as we know it today, was widely available by the 1940s. It completely changed human interaction with ink, the physical experience of writing, and, according to some, the prevalence of the ball-point pen killed elegant cursive handwriting.[6] With the introduction of new technology, there is always a gain and a loss. With the new technology of the ball-point pen, the beautiful cursive handwriting achieved with a quill or a fountain pen disappeared, yet we gained a relatively inexpensive pen that is conveniently transportable. The degree to which we gain or lose varies from one technology to another, but those losses and gains are an integral part of that process.

For those teachers who needed to manipulate images and show overlapping or moving geometric shapes, not possible on the chalkboard, the new technology of the overhead projected in the 1960s was a boon. The overhead projector projected the image of writing on clear film and was lit with a light source and projected onto a screen. Other new technologies that emerged post-World War II include headphones in 1950, which quickly found an educational use in foreign language lab instruction, among other areas of study. The handheld calculator (1972), and the Scantron system of testing (1972),

---

[5] G. Siemens, Massive Open Online Courses: Innovation in Education. In R. McGreal, W. Kinuthai, & S. Marshall (Eds.), *Open Educational Resources: Innovation, Research and Practice.* Vancouver: Commonwealth of Learning and Athabasca University, pp. 5–16.

[6] In the United States, John Loud received the first patent for the ball-point pen in 1888, however, it had problems with ink flow and was not successful. J. Giesbrecht, How the Ballpoint Pen Killed Cursive. *The Atlantic.* https://www.theatlantic.com/technology/archive/2015/08/ballpoint-pens-object-lesson-history-handwriting/402205/.

which allowed teachers to grade a large number of tests more quickly and efficiently, were immediate-response technologies that became necessary for the quick production and review of educational materials. By 1959, for example, using a photocopier was standard operating procedure in school systems across the nation.[7]

Entering more familiar territory, perhaps, is the personal computer, which is practically everywhere in some form from classrooms, offices, and homes to cell phones and wrist watches. The first computers were developed in the 1930s; however, computers for everyday use in homes and offices were not common until the 1980s. International Business Machines (IBM) introduced its first personal computer in 1981 to great fanfare. The following year, *Time* magazine named IBM's computer "Man of the Year." Toshiba and Apple quickly followed IBM with their own models—Apple's Mac was available starting in 1984, for example. In 1990, the World Wide Web was made possible through Hyper Text Markup Language (HTML), and three years later, when the National Science Foundation removed restrictions on the commercial use of the Internet,[8] a global communication system opened up new avenues of information sharing and research. By 2009, it was reported that 97% of classrooms in the United States had one or more computers.[9] In 2015, it was found that 94% of children ages 3–18 in the United States had a computer at home, and 61% of children ages 3–18 had Internet access at home.[10] The availability of the Internet facilitated social media. Social media platforms emerged in 2003 with MySpace, Facebook in 2004, Twitter in 2007, and Instagram in 2010. Instant connectivity is now the norm, and it is expected and demanded in both the education and business worlds.

---

[7] A History of Classroom Technology: The Primitive Classroom. Purdue Education Blog, Purdue University. Retrieved on June 01, 2020 at: https://online.purdue.edu/blog/education/evolution-technology-classroom.

[8] *Ibid.*

[9] U.S. Department of Education, National Center for Education Statistics, "Fast Facts." Retrieved on June 1, 2020 at: https://nces.ed.gov/fastfacts.

[10] *Ibid.*

All of this new technology was addressing a very real need in education in the United States, especially during the 20th century. According to the Department of Education, the high-school-enrollment rate for students 5–19-years-old rose from 51% in 1900 to 75% in 1940. By 1991, the high-school-enrollment rate was 93%. Not surprisingly, the dramatic increase in the number of students enrolling in college followed a similar trajectory to that of high school students.[11] With an ever-increasing student enrollment, educators adapted new or existing technologies for the specific needs of a rapidly expanding and changing educational environment.

As we move forward, changes and advances in technology seem to move at an ever-faster rate. New technology is introduced only to be obsolete soon thereafter. In the time it will take for this book to hit the market, technologies will have changed, which is why we avoid mentioning or recommending specific educational technologies as part of this discussion. The overarching theme in this chapter is that technology is constantly changing; therefore, teachers must be flexible and remain open to the potential of technology while, at the same time, rigorously evaluating whether a certain technology is necessary in the classroom. This leads to the three main areas of discussion in this chapter on technology and its successful integration into the classroom: First, the importance of teachers' willingness to adapt to change in the classroom. Change can be intimidating and threatening, especially when there is so much at stake for both the teacher and the student. Being open to change is not always easy. Second, the importance of teacher involvement in the choosing and introduction of new technologies in the classroom, and third, the importance of understanding whether any particular technology is truly needed in the classroom. This implies that one must be concerned that the introduction of new technology is supportive and not distracting, which can often be irrelevant to the producer of this new technology. How does the new technology enhance instruction? Of course, learning should

---

[11] T.D. Snyder (Ed.), *120 Years of American Education: A Statistical Portrait*. Department of Education, Office of Educational Research and Improvement, National Center for Education Statistics. January 1993.

be the primary impetus that drives the use of technology in the classroom. For example, when teaching geometry, one could use overlapping transparencies to demonstrate and identify congruent triangles more easily. Also, when one wants to demonstrate concurrency or collinearity of geometric figures, placing overlapping transparencies one by one introducing the components independently can have a great effect on student's learning.

If there is one thing we can all count on, it is that throughout life we will face change. Whether we adapt to these changes or not will determine how well we will do in our career. And no matter what our decision may be, the changes will still take place. We all know someone who has not adapted to change, such as a family member who refuses to use social media or email or cell phones. Yet, when the COVID-19 pandemic in 2020 made it necessary to close schools across the nation and globally, those who were not familiar with certain technologies such as Blackboard, Google Meet, and Zoom, among others, found themselves struggling to effectively reach and motivate their students through online teaching. Even those teachers who were familiar with online instruction were challenged because of the speed with which they had to make the adjustment from in-class to online teaching. Because they were familiar with the existing technologies, those teachers were better able to handle those additional challenges. While the pandemic was an unusual occurrence, the event dramatically demonstrated the importance of keeping up-to-date with educational technologies. However, up-to-date does not necessarily mean being an expert on these platforms.

Staying conversant with online educational technologies can mean "being familiar with" the technology. Perhaps you only know a small fraction of what educational technology offers in the way of instruction. That is a start! At least you know the product exists and what it can do. You can join the conversation, ask questions, and learn more about it. Being curious, flexible, and willing to adapt is a smart approach to learning that will support your development and growth as a teacher. Many professions require lifelong learning (also known as professional development) in order to keep up with changes— changes in technology, changes in laws, changes in policies, and changes in concepts and language that, at one time, seemed immutable (more on gender and language in Chapter 10).

Change can be scary. It can be intimidating and threatening, especially once you start to see yourself as a seasoned professional and are comfortable with how you currently teach. You have a stack of lesson plans that seem to work well. You have your standard examples or stories that effectively illustrate an important concept. You have recently learned how to use the overhead projector and even created beautiful transparencies. However, your principal may have recently received special funding that allowed the school to buy a computer for each classroom. Now all teachers are required to start using the personal computer in their classrooms. Overhead projector to computer—that was a big leap for educators at the time. The fear of change can be overwhelming, and most adults do not accept change easily.

In order to deal with the turmoil and frustration of change, teachers must be highly motivated to learn about and use new technology in the classroom. It also helps to have good models to emulate the effective integration of technology into the curriculum. Students are expected to struggle, stumble, and even fail while learning, and teachers provide the support and guidance for students as they go through this process. When integrating new technology into the curriculum, teachers need the support of administrators while they gain familiarity and comfort with the new technology that is to be introduced to their students. In addition, it is extremely important that teachers have a voice in the selection of new technology, as they are ultimately responsible for the learning that takes place in their classrooms. The skills and the attitude of the teacher is the strongest determinant in the effective integration of technology into the curriculum.[12]

Discussions around choosing and integrating new technologies into the classroom need to include the question about whether a new technology is actually needed. Putting budgetary issues aside for a

---

[12] N. Bitner & J. Bitner, Integrating Technology into the Classroom: Eight Keys to Success. *Journal of Technology and Teacher Education*, 10(1), 95–100 (2002).

moment—because if you decide you do need the new technology, securing the funding is the first consideration—just because the technology is available, is it the best approach for your students? Students may seem more engaged with that cool new device, but does that new device improve their understand of the topic? Teachers must take into account their particular student population, their various needs and challenges, so that the technology they chose to introduce into the classroom enhances rather than distracts from instruction. We can see this with the introduction of the chalkboard as a teaching technology in the mid-1800s since the chalkboard enhanced what teachers were already doing in the classroom.

There are certain geometric concepts which can be demonstrated with dynamic geometry programs that cannot be introduced to students in any way remotely as effective as the technology will provide. For example, suppose you want to introduce the theorem that if you connect the consecutive midpoints of any shaped quadrilateral, you will always end up with a parallelogram. You can then further impress the class as to by showing them what aspects of the original quadrilateral will yield a rectangle, a rhombus, and a square, rather than merely a general parallelogram. Using dynamic geometry software, have each of your students draw an ugly (i.e. any shaped) quadrilateral. Then have them locate the midpoints of the four sides of the quadrilateral. Now have them join these points consecutively. Everyone's drawing should have resulted in a parallelogram. Wow! How did this happen? Everyone began (most likely) with a different shaped quadrilateral. Yet everyone ended up with a parallelogram. Up until this point, the normal paper and pencil would have sufficed. However, the computer program now allows each student to distort the original quadrilateral into a different shape while at the same time noticing that the resulting quadrilateral formed by joining the midpoints will always remain a parallelogram. Figure 7.1 shows a few possible results:

A question that ought to be asked at this point is how might the original quadrilateral have been shaped for the parallelogram to be a rectangle, rhombus or square? Once again, by manipulating the original quadrilateral, students should notice the following:

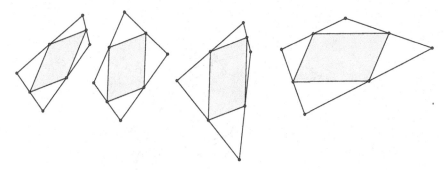

Figure 7.1

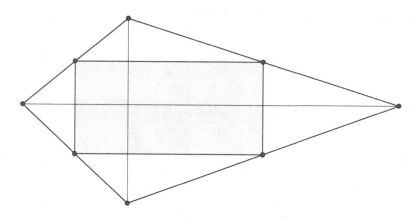

Figure 7.2

When the diagonals of the original quadrilateral are perpendicular, the parallelogram is a **rectangle**, as we see in Figure 7.2.

When the diagonals of the original quadrilateral are congruent, then the parallelogram is a **rhombus**, as we can see in Figure 7.3.

When the diagonals of the original quadrilateral are congruent and perpendicular, then the parallelogram is a **square**, as shown in Figure 7.4.

Even though the dynamic geometry software should be sufficiently convincing, to *prove* that all of the above is "really true," a short proof outline is provided, one that should be in easy reach for a high school geometry student.

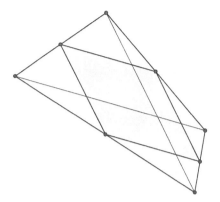

Figure 7.3

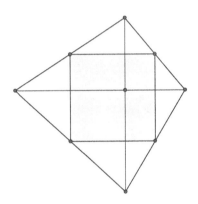

Figure 7.4

Proof Outline

The proof is based on a simple theorem that states that a line segment joining the midpoints of two sides of a triangle is parallel to and half the length of the third side of the triangle. This is precisely what happens here, as we will see in our proof using Figure 7.5.

In $\triangle ADB$, the midpoints of sides AD and AB are F and G, respectively. Therefore, $FG \parallel DB$ and $FG = \frac{1}{2}BD$, and $EH \parallel DB$ and $EH = \frac{1}{2}BD$.

Therefore, $FG \parallel EH$ and $FG = EH$. This establishes $FGHE$ a parallelogram.

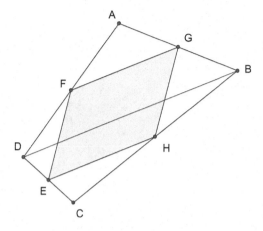

Figure 7.5

Furthermore, if the diagonals *DB* and *AC*, are congruent, then the sides of the parallelogram must also be congruent, since they are each one-half the length of the diagonals of the original quadrilateral. This results in a rhombus.

Similarly, if the diagonals of the original quadrilateral are perpendicular and congruent, then since the sides of the parallelogram are, in pairs, parallel to the diagonals and half their length, the sides of the parallelogram must be perpendicular and congruent to each other, making it a square.

This discussion clearly demonstrates for students a geometric phenomenon that they can actually witness through the various distortions of the original quadrilateral to demonstrate that it should hold true for all quadrilaterals. Naturally, a proof is still in order.

Chapter 8

Navigating School Culture

Any visitor to a school will immediately get an important insight into that school's culture simply by walking through the front door. A visitor may be greeted by friendly security personnel who ask, with a smile, "How are you today?" before requesting identification, printing an identity badge, and providing clear directions to the visitor's destination. Or, one may be greeted with a sign in big red letters that says "Stop!" or "Warning!" with a message letting the visitor know that they should not proceed without permission. Maybe, eventually, a staff member wanders out to see what the visitor wants. The first school has a culture that welcomes not only the students, staff, and teachers but also parents and community stakeholders. The second school, unfortunately, does not value the experience of feeling welcomed and respected, and they are sending a strong message about the culture of their school. These initial impressions set the tone for whatever meetings might take place at these schools as well as the daily teaching and learning that go on there. How does one school, with the warm welcome, come to that culture of caring, while the other school, with the big "stop sign" at their entrance, seems not to care? The answer lies in the complex interactions among administrators, teachers, staff, students, parents, and community stakeholders as well as how their shared purpose and goals are communicated and acted upon. Core beliefs and behaviors are the building blocks of any school culture.

As a new teacher, trying to figure out and navigate a school's culture can be tricky. Before you start to think about navigation strategies, possessing a strong sense of what the culture is at a particular school is critical. A solid understanding will help you determine whether you are a good fit for the school and if the school will be a good fit with your beliefs, values, and career goals in education. When you interview for a teaching job at a new school, it is rare that someone says: "I'd like to explain what this school's culture is." Yet, it might be one of the most important things to know about the place where you will spend the majority of your waking time. First, if you are thinking about what kind of environment you will be entering— the values and beliefs of the place and people—before your interview, you can directly ask the search committee about the school's culture. Be prepared to explain to the committee *exactly* what you mean by "school culture," which we explore in this chapter. Second, remember that once you accept a teaching position at a school, it is highly unlikely that you will be able to change the culture of that school.

In this chapter, we will examine school culture as organizational culture. An organization is a body of people with a particular and shared purpose. Often when people mention an organization, they are referring to a company, an institution, or an association. Education is one of the most basic institutions in the United States, and each school district and each school will have its own unique culture— unlike most businesses, which rely on a uniform and and consistent use of a brand identity, regionally, nationally, or globally. Federal and state governments and, to a large extent, school districts determine much of education policy as well as curriculum, performance goals, outcomes, and assessment. These mandates are passed down from government entities to individual school districts. How each school reaches those student achievement goals or meets policy guidelines varies widely—and relies heavily on the ability to perform. Those who study organizations have found that there is a strong and significant correlation between an organization's culture and that organization's performance.[1] For those education leaders interested in student and teacher performance, building a positive culture is key because organizational culture is a strong determinant in a school's success. Though hard to pinpoint, because "culture" seems to denote a slew of slippery intangibles, we can actually identify some characteristics of strong school cultures and weak school cultures.

A strong school culture results from a clearly identified and widely communicated mission or vision.[2] In a setting that has a strong school culture, teachers share this mission/vision with leadership and, as a result, a majority of the students are provided a platform where success is more likely. A weak school culture will have low morale and rapid turnover in staff and teachers. Students are most likely struggling in this environment because it lacks consistency and focus. According to Dr. Ebony Bridwell-Mitchell, at the Harvard Graduate School of Education, "[a] culture will be strong or weak depending on the interactions of people in the organization. In a strong culture, there are many, and overlapping, cohesive interactions, so that knowledge about that organization's distinctive character—and what it takes to thrive in it—is widely spread. In a weak culture, sparse interactions make it difficult for people to learn the organization's culture, so its character is barely noticeable and the commitment to it is scarce or sporadic."[3] The quality and frequency of communication and interaction among staff and teachers, therefore, are a few of the elements that will determine whether a school's culture is strong, weak, or just mediocre.

---

[1] See D.R. Denison, & A.K. Mishra, Toward a Theory of Organizational Culture and Effectiveness. *Organization Science*, 6(2), 204–223 (1995); Z. Aycan, R.N. Kanungo, & J.B.P. Sinha, Organizational Culture and Human Resource Management Practices: The Model of Culture Fit. *Journal of Cross-Cultural Psychology*, 30(4), 501–526 (1999).

[2] "Mission" and "vision" are sometimes used interchangeably, yet some make the distinction that the mission carries out the vision. In this case, two separate statements are created: one for vision and one for mission. It is also possible that a mission statement may incorporate the vision aspect into the mission statement and vice versa.

[3] Leah Shafer, What Makes a Good School Culture. *Usable Knowledge: Relevant Research for Today's Educator.* Harvard Graduate School of Education, July 23, 2018.

The messages that leaders of any school environment send, whether direct or indirect, also contribute to the culture. This becomes apparent in the chapter's opening paragraph examples of the warm welcome a visitor receives from a school with a strong culture, and the lack of a welcome from a school with a weak culture. These are messages communicated from leadership: indirectly, through the effort dedicated to a warm welcome, or, conversely, through the lack of personal contact or some other means to greet visitors. School leaders can also directly communicate the school culture through the creation of a clearly written school mission/vision statement. The mission/vision statement should be prominently displayed, widely circulated, and referred to often in staff and teacher meetings as well as school assemblies, when appropriate.

The mission/vision statement should be treated as a "living" document in that it is not a static announcement of what the school intends to accomplish and how. Rather, a cleverly executed mission/vision statement leaves room for revisions and updates. For instance, a school may need to restate or update its goals. Or a school discovers that a particular process is not working or a policy has changed, so the school updates its mission/vision statement to reflect those new developments. If you are fortunate, the school at which you are interviewing might actually have a mission/vision statement on its website. An example of a mission/vision statement might look something like this: *Our mission is to provide high-quality education in a safe, respectful, and inclusive environment that builds a foundation for lifelong learning. We strive to create this environment by building a more inclusive curriculum, addressing a range of learning styles, offering an array of diversity programs, and by participating in the wider community.* This statement has a strong focus on inclusivity, so one would expect to see that reflected in the school culture. Make sure to scour the website of the school district as well as the individual school for mission/vision statements and comparably relevant information.

A mission/vision statement in the education world is actually quite similar to product branding in the business world. The mission/vision of a business will determine the development and management of a brand. Businesses strive for consistency when building a product brand in terms of product name, product design, brand story, marketing, and intended consumer experience. In terms of brand recognition and consistency, for example, the athletic-wear company Nike is probably one of the most recognizable and well-managed brands in the world. The Nike name and trademark "swoosh" appear in the same font and design in all marketing campaigns. The Nike company spends a considerable amount of resources protecting the brand, requiring careful attention to consistent design and presentation. This attention to brand image certainly extends to the company's selection of celebrity and athlete endorsements.[4] Some brands are so well-known that consumers can identify them simply by a silhouette or the shape of its container such as the curved glass bottle of the Coca-Cola soft drink brand. However, it is also imperative that brands remain flexible so that they can change according to consumer demand, when there is a slump in product revenue, and when there are social changes that impact the brand. We see similar adjustments in education when curriculum changes are necessary to reflect a diverse population or a historical perspective that had been previously ignored, for example, or when new afterschool programs are added to serve student needs. These changes should be celebrated and communicated by school leaders as a positive adjustment made for the benefit of students, teacher, parents, and other stakeholders.

Even when a company's product is a service—for example, legal services, beauty services, or hospitality services—companies still spend major resources to build that brand. Marriott International is a good example of a company that has a clearly stated core value of "putting people first," which is consistently communicated as the central mission. An important part of the brand story of this hotel chain is the history of Marriott and the founders, J. Willard Marriott and Alice Sheets Marriott. This husband and wife team had a simple philosophy: "Take care of associates and they will take care of the customers." This reflects their value of putting people first. The company mission and vision statement also emphasizes that Marriott pursues

---

[4] Brands may change a name or design when entering a market in another country in acknowledgement of local cultural values and practices.

excellence (superior customer service), embraces change (innovation and growth), acts with integrity (ethical and legal standards), and serves our world (sustainability and social impact).[5] This is a top-notch mission/vision statement.

Similar to public relations in business, schools must present themselves to not only the district, state, and federal leaders, but they must also make a good impression on relevant stakeholders: parents, community leaders, funders, corporate partners, local government officials, and others, depending on that particular school. A savvy education administration might borrow some techniques from the business world to make sure they are presenting their achievements in a clear and consistent manner through a mission/vision statement, annual report, and/or newsletter. They should strive to capture a wide audience in their communications (students, parents, teachers, staff), and consistently reach out to all stakeholders.

But, unlike businesses that have a narrow focus on the consumer and profits, school leaders have a variety of constituents to keep in mind, so they actually have to be smarter about the ways in which they communicate who they are and what they have accomplished (brand identity). When you think about what is at stake in education, the importance of school public relations is obviously vital.

Clearly communicating and promoting a strong school culture (the school's brand) is critical to a school's success. Leaders can promote a strong culture by sending messages that encourage traits such as collaboration, transparency, and the value of hard work. Shafer identifies five interconnected elements of culture, that can be influenced by school leaders, to create a strong school culture[6]:

- **Fundamental beliefs and assumptions:** The school's belief, for example, that "all students have the potential to succeed." This may be included in the mission/vision statement.

---

[5] www.Marriott.com.
[6] Leah Shafer, What Makes a Good School Culture. *Usable Knowledge: Relevant Research for Today's Educator.* Harvard Graduate School of Education, July 23, 2018.

- **Shared values:** These are the judgements that stakeholders (leaders, staff, teachers, etc.) at your school make about those beliefs and assumptions. "The right thing is for all teachers to be collaborating with each other to promote student success," for example.
- **Norms:** How all stakeholders believe they *should* act and behave, what they believe is expected of them. "We should greet all students and visitors to our school with a warm welcome," or "We should be fully engaged at staff and teacher meetings."
- **Patterns and behaviors:** This is the way that leaders, teachers, and staff *actually act* at the school. These patterns and behaviors should reflect the norms. For example, do visitors actually get a warm welcome at the school? Are staff and teachers fully engaged at weekly meetings? Or is this just wishful thinking? In a weak school culture, the patterns and behaviors do not match the norms.
- **Tangible evidence:** These are the physical, visual, auditory, and other sensory signs that reflect the behavior of the leaders, teachers, and staff in the school. An example would be prominently displayed student artwork in the hallways, a television monitor that announces student achievements for all to view, a school principal and rotating group of teachers who stand at the entrance every morning greeting students and parents.

Each of these five elements builds upon the others and forms a supportive framework for interactions among all stakeholders of a school. Strong connections and clear communication among the stakeholders also reinforce these elements.

In particular, the last of these five points provides mathematics teachers with a variety of options to popularize the school's achievements and the student's enthusiasm. For example, at a strategically placed position a mathematics bulletin board that exhibits outstanding student works there could also be a "Problem of the Week" that would be presented each week as a challenge to students, with the successful students' submissions listed the following week along with one of the more impressive solutions. Mathematics teachers also strengthen the culture of a school by conducting extracurricular programs that can enhance the posture of mathematics for students as well as the greater community. For example, to put on an assembly program showing unusual aspects of mathematics to both students and parents. Such things as showing what a Mobius strip is and its properties when it is cut halfway from an edge or one quarter the distance from an edge. The results are typically astonishing to a general audience. There are countless other arithmetic, geometric, and probability wonders that are clearly not a part of a secondary school curriculum and yet can be easily understood by the general audience of both students and parents.

Just to whet the readers' appetite, we will provide an example in each of these three previously-mentioned categories: an example in arithmetic would be to show how you can determine whether a number is divisible by 3 or 9 by simple inspection—that is, when the sum of the digits is divisible by 3 or 9. Geometry lends itself to a plethora of examples, we will just mention one known as Napoleon's theorem, although it is disputed whether Napoleon actually discovered this relationship or if it was one of his supportive military engineers.

The theorem is based on drawing and equilateral triangle on each side of any shaped triangle, as we show in Figure 8.1. The theorem states that *the segments joining each vertex of a given triangle (of any shape) with the remote vertex of the equilateral triangle (drawn externally on the opposite side of the given triangle) are congruent and concurrent.*

That is, *AE, BD,* and *CF* are congruent to one another. Take note of the unusualness of this situation, since we can start off with *any* shaped triangle and still this relationship holds true. If you were to draw your own original triangle you will come up with the same conclusion. Either straightedge and compasses or dynamic software programs such as *Geometer's Sketchpad* or *GeoGebra* would be more dramatic in showing this amazing relationship.

Before we embark on the adventures that this theorem provides, it may be helpful to offer a hint as to how to prove this theorem. The trick is to identify the appropriate triangles to prove congruent. They are not too easy to identify. One pair of these congruent triangles is

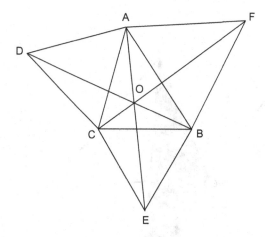

Figure 8.1

shown in Figure 8.2 and can be used to establish the congruence of *AE* and *BD*. The other segments can be proved congruent in a similar way with another analogous pair of congruent triangles embedded in the figure.

There are quite a few unusual properties in this configuration. For example, you may not have noticed that the three segments *AE, BD*, and *CF* are also concurrent. This concept is not much explored in the typical high school geometry course. For our purposes here we shall accept it without proof.[7]

Not only is point *O* a common point for the three segments, but it is also the only point in the triangle where the sum of the distances to the vertices of the original triangle is a minimum. This is often called the *minimum-distance point* of the triangle *ABC*. The sum of the distances from any other point in the triangle to the three vertices would be greater than from this minimum-distance point.

---

[7] For a proof of this theorem and its extensions see A. S. Posamentier, *Advanced Euclidean Geometry: Excursions for Secondary Teachers and Students,* Chapter 4. Emeryville, CA: Key College Publishing, 2002. Also see A. S. Posamentier & I. Lehmann, *The Secrets of Triangles: A Mathematical Journey.* Guilford, CT: Prometheus Books, 2012, pp. 66–75.

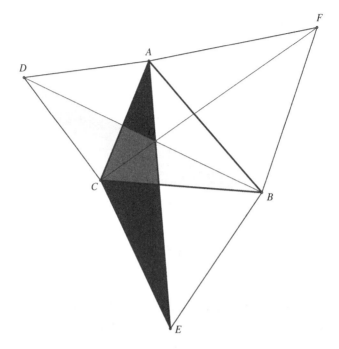

Figure 8.2

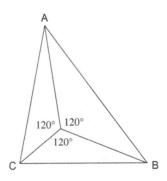

Figure 8.3

As if this weren't enough, this point of concurrency is the only point in the triangle where the sides subtend equal angles. That is, $\angle AOC = \angle COB = \angle BOA = 120°$ (see Figure 8.3).

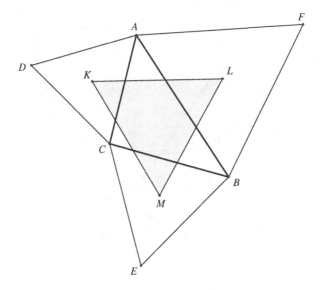

Figure 8.4

There is more, and that is why this would be an illustrative example of entertaining a general audience of students and parents. Moving along, we locate the center of the each of the three equilateral triangles, which can be done in a variety of ways: find the point of intersection of the three altitudes, medians, or angle bisectors. Joining these center points reveals that another equilateral triangle has been created, as we show in Figure 8.4. Remind your audience that we began with just any randomly drawn triangle *ABC* and now all of these lovely properties appear.

With a dynamic software program such as *Geometer's Sketchpad* or *GeoGebra* you can see that regardless of the shape of the original triangle the above relationships all hold true. To enhance the presentation one might present an interesting question such as what would we expect would happen if point *C* were to be on *AB*, thereby collapsing the original triangle, as we show in Figure 8.5.

Lo and behold, we still have our latest equilateral triangle preserved. Perhaps even more astonishing (if anything could be) is the generalization of this theorem. That is, suppose we were to construct

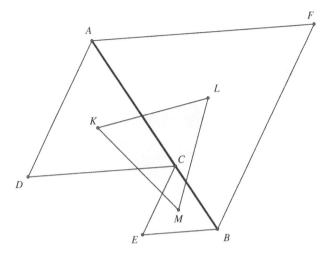

Figure 8.5

similar triangles appropriately placed on the sides of our randomly drawn triangle, and joined their centers (this time we must be consistent as to which center we choose to use—centroid, orthocenter, or incenter[8]), the resulting figure will be similar to the three similar triangles.

With the aid of a dynamic software program we can see that all that we said above about triangles drawn *externally* on the sides of our randomly selected triangle can be extended to triangles drawn *internally* as well. So, from this simple theorem came a host of incredible relationships, many of which can be discovered independently— in particular, if you have a dynamic software program available. Thus, we can see how geometry can be used to surely entertain an audience and thereby enhancing the culture of the school.

There are lots of types of averages, which in various ways measure a point of central tendency. There is the usual average, known in

---

[8] The *centroid* is the point of intersection of the medians of a triangle. The *orthocenter* is the point of intersection of the altitudes of a triangle. The *incenter* is the point of intersection of the angle bisectors of a triangle.

mathematics as the arithmetic mean. Then there is a geometric mean, and a harmonic mean. However, there are other measures of central tendency, such as the mode of the set of data, which refers to that item or score that comes up most frequently. Then there is the median, which is the midpoint of a spectrum of data, regardless of where the strong points or weak points are located. We need to dispel the use of the word average as it has been used so frequently in sports, in particular, in baseball.

Although baseball batting averages seem to permeate sports discussions, few people realize that these "baseball batting averages" aren't really averages in the true sense of the term, they are actually percentages. Most people, especially after trying to explain this concept, will begin to realize that it is not an average in the way they usually define an "average"—the arithmetic mean. It might be good to search the sports section of the local newspaper to find two baseball players who currently have the same batting average, but who have achieved their respective averages with a different number of hits. We shall use a hypothetical example here.

Consider two players: David and Lisa, each with a batting average of 0.667. David achieved his batting average by getting 20 hits for 30 at bats, while Lisa achieved her batting average by getting 2 hits for 3 at bats.

On the next day both performed equally, getting 1 hit for 2 at bats (for a 0.500 batting average); one might expect that they then still have the same batting average at the end of the day. Calculating their respective averages: David now has $20 + 1 = 21$ hits for $30 + 2 = 32$ at bats for a $\frac{21}{32} = 0.656$ batting average.

Lisa now has $2 + 1 = 3$ hits for $3 + 2 = 5$ at bats for a $\frac{3}{5} = .600$ batting average. Surprise! They do not have equal batting averages.

Suppose we consider the next day, where Lisa performs considerably better than David does. Lisa gets 2 hits for 3 at bats, while David gets 1 hit for 3 at bats. We shall now calculate their respective averages:

David has $21 + 1 = 22$ hits for $32 + 3 = 35$ at bats for a batting average of $\frac{22}{35} = .629$.

Lisa has 3 + 2 = 5 hits for 5 + 3 = 8 at bats for a batting average of $\frac{5}{8} = .625$.

Amazingly, despite Lisa's superior performance on this day, her batting average, which was the same as David's at the start, is still lower. There is much to be learned from this "misuse" of the word "average," but more importantly one can get an appreciation of the notion of varying weights of items being averaged.

Now that you understand school culture and what determines whether a school has a strong or weak culture, consider how interpersonal relationships contribute to a school's culture and the role that you should play in these interactions. We will focus mainly on your relationships with colleagues and with the leaders of the school. These interpersonal relationships are sometimes called "office politics." But this term often has a negative connotation. First, to use the word "politics" brings up the typical government two-party (or more) politics, which is not what we focus on here. Second, "office politics" connotes a negative string of interactions that often include back-stabbing, Machiavellian plotting, and teacher's lounge gossip. As a professional, of course, you should avoid these negative interactions because they have a way of coming back to haunt you. Whether it is a seemingly harmless joke about your supervisor or repeating gossip about a colleague—you might be surprised by how quickly these kinds of remarks spread through the school gossip vine. This is the fastest way to earn a bad reputation.

Just as in any work environment, complicated personal dynamics in a school environment do exist. New teachers should learn how to deal with certain interpersonal situations and to navigate the delicate complexity of how people interact. In a strong school culture, the vast majority of your colleagues should be focused on the mission/vision, and their patterns and behaviors (what *actions* they actually take to support the mission/vision) should reflect that. Perhaps you find yourself in a school with a weak or mediocre culture. Because collaboration and communication are probably not shared values in a weak school culture, you might find it difficult to break into the tight-knit group of colleagues who teach your grade level. Naturally, it is the teachers who teach the same grade level who often band together—eating lunch together, sitting together at meetings, taking their respective classes on field trips together, or meeting parents together. There could be benefits to being a part of this group, but depending on the reputation of this group, you may not want to join them. Even if you do, it can be a challenge to break into these tightly formed inner-circles.

Complicated group dynamics also occur in the business world (and in most industries) where employees are divided into departments, groups, or teams. It is natural to build a bond with the people in your group, and it is actually critically important to have a strong working relationship with these team members. However, in order to avoid a silo mentality,[9] many forward-thinking businesses with a strong culture will encourage employees to share information and resources with other departments to benefit the whole. This can be a challenge when people in a particular department "guard" or feel that they "own" company information or resources. These five basic suggestions can help you navigate interpersonal communication and interactions beyond the classroom setting.[10]

1. Be aware of the particular personnel dynamics at your school (part of the school culture). When first starting your new position, spend a lot of time observing interactions between teachers and administrators, between staff and teachers and parents. Understand the dynamics that exist.

2. Remain outside of those personnel dynamics that may negatively affect your position or reputation at the school. Once you become associated with a particular group, what they represent and how

---

[9] Organizational silos (a department, team, group, etc.) can prevent information and resources from being shared among departments and teams. If collaboration and team-building are not encouraged, silo mentality can have a negative impact on business.

[10] Steve Haberlin, The Politics of Teaching. *Education World*. Retrieved on June 20, 2020 at: https://www.educationworld.com/blog/politics-teaching.

they act will reflect on you—whether or not you are actively engaged with their agenda or just "hanging out" with them.

3. Be a professional and treat *everyone* with respect. Everyone means not just the leaders and your colleagues, but includes the office staff, janitors, and cafeteria staff. Work on developing a strong and positive reputation at your school and understand that what you do and say has a ripple effect that may impact your reputation.

4. Use emotional intelligence[11] when dealing with colleagues and leaders—this is instrumental in becoming a professional. When you allow emotional intelligence to guide your responses to people and situations, you are using the rational part of your brain. This rational and considered response is in contrast to the more emotional and intuitive part of your brain that *reacts quickly.* There is certainly a time and place for a reactionary response; however, the workplace is not usually one of them! Turn off the "default" part of your brain that reacts without much thought, and, instead, operate using your emotional intelligence and respond rationally.

5. Always go back to the mission/vision. If, by chance, you are not immediately accepted into the click-ish inner-circle of teachers, stay focused on why you are there. Work hard for your students, stay positive, and stay above the fray.

These five points are relevant for any work situation, yet, as mentioned earlier in this chapter, teachers directly interact with more stakeholders than do most employees in the business world or other industries. Keep these ideas in mind when dealing with colleagues, leaders at your school and beyond, including school board members, parents, and other stakeholders such as funders, corporate partners, and community leaders.

---

[11] Emotional intelligence is the capacity to be aware of, control, and express emotions, and to handle relationships judiciously and empathetically. Dictionary.com.

Now you should understand the basics of organizational culture—strong and weak—and know that leadership plays a critical role in the development of a culture. So, what exactly does this leadership look like? A leader in a school with strong culture will be a collaborator and hold high expectations for teachers, staff, and students. They will model the beliefs and behavior outlined in the mission/vision and engage in honest conversations about the attitudes, norms, and practices that are important to the school's culture. They inspire and reward excellence. This can be enhanced by employing the extracurricular activities that a mathematics teacher can provide for the school and beyond and that will allow the school to most likely standout favorably from others.

In the next chapter, which focuses on leadership, you will read about the various types of leaders and how to deal with the particular characteristics of each type. The successful leader who has managed to build a strong school culture will undoubtedly use a combination of leadership approaches depending on who they are dealing with and the situation at hand.

Chapter 9

Developing Leadership Skills

Knowing yourself—your strengths and weaknesses—is an important element of successful teaching (as discussed in Chapter 1). These insights into personality help you to develop your skills as a teacher and, upon deeper reflection, might also reveal any leadership tendencies you may already possess. The field of education needs strong leaders, and as you start your career, you may want to consider the possibility of taking on a leadership role outside of the classroom at some time in the future. But start asking yourself early on in your career: What does effective leadership look like in education? Who are the role models to emulate? When you understand the characteristics of a good leader, you can incorporate some of those qualities into your own classroom leadership style. These skills will help you as a leader in the classroom and can potentially transfer to projects outside of the classroom that can advance your career in education. In addition, a familiarity with the different types of leadership styles will enable you to understand your supervisors and other leaders in your school district, which ultimately gives you an edge in those important work relationships.

In the world of business, the successful leaders who are mentioned time and time again include former and current CEOs Andrew Grove (Intel), Sheryl Sandberg (Facebook), Jack Welch (General Electric), Kenneth Chenault (American Express), and Jeff Bezos (Amazon), among others. These CEOs represent very different leadership styles, but they all have one thing in common: Through innovative thinking and action, they led their companies to realize extraordinary profits. Welch, for example, catapulted GE into the profit stratosphere by adding healthcare, venture capital, and financial services[1] to a company that had been known for lighting and power generation. The study of leadership in business management practically overshadows all other topics because the industry recognizes that without good leaders, profits suffer and businesses do not survive. This is evidenced by a plethora of business books and articles that focus on effective leadership, qualities of an effective leader, how to develop strong leadership skills, and trends in leadership.

The field of education, however, does not have the same goals as business in that we are not focused on profits. The great leaders in education have often been reformers such as Booker T. Washington, who fought for civil rights and desegregation in schools; Horace Mann, who developed the American public-school system; Margaret Bancroft, who created the field of special education, and John Dewey, who was an advocate for progressive education. Geoffrey Canada, President and former CEO of The Harlem Children's Zone, is the best example of a modern-day reformer. His visionary work helping children and families who live in poverty in Harlem is world-renowned, and he is a tireless advocate for education reform. *Fortune* Magazine named Canada #12 on the list of "World's 50 Greatest Leaders" in 2014.[2] These leaders in education were (and in the case of Canada— are) focused on improving conditions for students, and, as a result, contributed to building a stronger and more just society in which a greater number of people could realize their potential. Even though our goals in education may be different from the goals of business, we might want to cultivate the same qualities in our leaders as a corporation, for instance, seeks to develop in their managers and CEOs. And there may be some interesting overlap or connection between the

---

[1] Although some are of the opinion that this diversification ultimately led to problems years later for the company.
[2] Retrieved on June 22, 2020 from: https://hcz.org/news/fortune-names-geoffrey-canada-list-worlds-50-greatest-leaders/.

business world and the field of education regarding effective leadership. Many of the qualities of great leaders transcend industry boundaries. In fact, some people believe that the best leaders are also great
teachers.

In his *Harvard Business Review* article, "The Best Leaders Are
Great Teachers," Sydney Finkelstein, professor of management at
Dartmouth College, says that "teaching is not merely an 'extra' for
good managers; it's an integral responsibility. If you're not teaching,
you're not leading."[3] Finkelstein describes a style of leadership that is
able to zero in on what information or lesson is important to teach,
the appropriate time to teach, and how to make those lessons stick.
One of the approaches these superstar teacher–leaders take is to
spend intensive one-on-one time tutoring their direct reports.
Personalized instruction of this kind has long been recognized by
teachers, psychologists, and education consultants, but it is actually a
rarity in the business world. When this leadership teaching does
occur in a business environment, the results are positive. Perhaps this
is not so surprising to those who work in education; however, it does
illustrate that certain approaches or strategies between business and
education can be shared with success. Business can learn from the
field of education that personalized instruction is valuable and the
benefits of teaching skills in management are measurable and have
the potential to add to the bottom line. Educators can explore the various leadership styles from business and adapt to fit certain classroom
and management situations.

Before we go further, it is important to outline and define the
main types of leadership commonly seen in an organizational or institutional setting—including both education and business. As you read
the descriptions of each leadership style, note which one you feel is
most closely aligned with your personality. You may even have some
experience having had a leadership role in a volunteer experience or
other position, and will be able to easily see yourself in one of these
descriptions. In addition, most people have a "default" leadership

---

[3] S. Finkelstein, The Best Leaders Are Great Teachers. *Harvard Business Review*, Jan/
Feb (2018).

style, which consists of the actions you would naturally default to when put in a leadership situation with little or no leadership experience. This default style is influenced by your personality type as well as by the type of leaders you have been exposed to during your life, including family members (see Chapter 1). Do not let your default leadership style (or tendencies) define who you are as a leader. Contrary to the old adage, leaders are *not born*, they are usually made through self-reflection and a lot of practice.[4] Keep in mind that a successful leader often combines aspects from a number of these leadership styles and adapts them to meet the needs of specific situations. In addition, each leadership type listed below exists on a continuum, so there could actually be an extreme example of an authoritarian leader as well as a leader who exhibits authoritarian *tendencies* on the opposite end of that spectrum.

Leadership Types

The *authoritarian* (or autocratic) leadership approach works best when students need strong direction from a leader—someone who is an expert on the topic. This sounds exactly like the traditional classroom teacher, right? Knowing your students is critical to your ability to use the appropriate leadership approach. When you are introducing a new concept to your students, this may be exactly the right approach. There are times when the authoritarian approach is not appropriate for the classroom setting, particular students, or a specific lesson. While some students may respond well to an authoritarian style of teaching, there may be others who rebel against that approach. Perhaps it is the student who is slightly advanced and does not need the firm guidance of what has been called the "Oz" model of leadership, which is a reference to the film, *The Wizard of Oz*, and the (seemingly) all-knowing and all-powerful wizard. With this student, you may need to find an approach that works with their specific needs. If your supervisor is the authoritarian leader, you will benefit

---

[4] We do not go into detail on the topic of leadership qualities in the classroom since that is covered in Chapter 4.

from their knowledge and may enjoy professional development opportunities. On the other hand, you may not have much say in decision-making processes. Develop the skills necessary to make this leader feel supported while maintaining your professional integrity. Not surprisingly, the authoritarian does not like to be challenged. If you do disagree with this all-knowing, all-powerful wizard or would like to offer an alternative to one of their ideas, make sure you know what you are talking about and choose your words carefully.

The *coaching* (or transformational) style of leadership focuses on helping the student learn and grow, much like a coach would provide guidance to an athlete to achieve certain goals. This approach only works with those who are receptive to learning and to change. Students who need a bit more incentive to learn would not do well with this style of leadership, for example. The coach–leader offers guidance and support, which could be great qualities in a supervisor of teachers. If your supervisor has the qualities of a coach–leader, take advantage of the opportunity to learn and grow under that guidance. Realize, however, that you must be motivated to grow. Unlike some of the other types of leaders who motivate using authority or even fear, the coach expects you to be self-motivated and focused on growth. A wonderful resource, specifically geared to teachers taking on a math coaching position, is the book *The Mathematics Coach Handbook*, by A.S. Posamentier and S. Krulik (World Scientific Publishing, 2019).

If someone is using fear to motivate or is strictly performance-oriented, you are most likely dealing with a *coercive* (or transactional) leader. This is the most directive of the leadership styles (i.e. "Do what I tell you right now"). Yet, as a teacher, you may need to use this approach if your classroom—or a student—is out of control. Coercive supervisors are often laser-focused on performance, so if you are dealing with this type of leader as your supervisor, make sure you know exactly what is expected of you so that you can do well on your annual review as well as in other areas of your work that are reviewed. You may not get much coaching from this leader, so look for a mentor beyond this particular supervisor if you feel that you could benefit from some professional guidance and support.

The *democratic* leadership approach can be useful in a classroom setting when you want to engage students. In this case, all students would have a say in a discussion, a project, or in a decision-making process. Democratic leaders provide an opportunity for all students to share ideas—so make sure you have the time to fully develop this type of discussion or project. A democratic approach may not work well with large classroom engagement, so arranging students in small groups might be one way to deal with the time issue. Along these lines, if your supervisor has a tendency toward democratic leadership, staff meetings might take a bit longer than, for instance, meetings run by an authoritarian leader. The democratic leader wants to make sure that everyone in the meeting is in agreement or at least has had their say. Consensus-building is time-consuming. Yet, employees feel much more appreciated and experience greater work satisfaction when they have a role in decision-making processes and feel their opinions are valued. It has been observed that when small groups are tasked to discuss an issue that would lead to change, very subtle suggestions by the leader often have a subconscious effect on the outcome of the small groups who tend to think that it is wise to conform to the leader's beliefs. Although this is not always the case, it is worthwhile to consider.

The *pace-setter* and *visionary* are very similar leadership types. Steve Jobs, the founder and former CEO of Apple, is often mentioned as an example of this type leader. Clearly, Jobs was a visionary—he was an innovator in the technology field and built a company that continues to innovate and grow by leaps and bounds. Many pace-setters or visionary leaders are "lead-by-example" types. They have very high expectations, demand quick results, and expect their employees to keep up with them. This kind of environment can be exciting and can inspire tremendous productivity. However, it can also lead to employee burn-out and sometimes even depress innovation, because the work is moving so quickly there is no time to reflect or adjust. If your supervisor fits into this pace-setting/visionary category, make sure you take care of your health so that you can maintain the physical and mental energy necessary to function in this environment. Short-term, this environment could be a springboard for advancement. If you have strong organizational skills, you might have the opportunity to work on multiple projects that might not be open to you with a supervisor who is not a pace-setter/visionary. Find a way to take advantage of these opportunities and make the experience work for you in terms of career and professional development. In the classroom setting, you can use the pace-setter/visionary leadership approach when you identify those students who can work at a faster pace than others. Find ways to challenge those students and keep expectations high for them. The pace-setter/visionary is always focused on the long-view. What is your vision for your students? Have in mind what you want your class to accomplish by the end of the year and maintain a steady pace for all students, while avoiding the burn-out that results from constant pressure (some pressure is good, too much pressure results in either burn-out or drop-out).

The *service* leader type has a "people-first" mindset and a desire to collaborate, build morale, and inspire their employees to find meaning and connection between their work and personal values. The service leader leads by serving and supporting others first. This leadership approach has the potential to work well in the classroom, although you might find that you need to incorporate some of the other leadership approaches mentioned here, especially if you have students who need extra incentives to get their work done. As a teacher, you may flourish working for someone who has the qualities of a service leader, especially if you share their sense of duty to serve. These shared values and beliefs are the foundation of mentoring and service more generally, which is how many teachers think of the work they do in the classroom that serves the greater good of society. If your ambitions are to advance quickly in your career, the service leader might not be interested in helping you get that big promotion—although this depends on the particular supervisor. A service leader would most likely feel that it is part of their duty to help their direct reports' advance.

The *laissez-faire* approach can hardly be considered leadership, can it? Actually, it can. A good example of a laissez-faire (the translation from French is "to let go") approach would be to let a group of students (or employees) function without much interference.

When used at the right time, and in the right situation, laissez-faire can be a great leadership approach. Imagine that you have presented new material to the class (authoritatively), and then followed that with an open discussion during which all students have the opportunity to ask questions and share ideas (democratically). You may then want to create small groups and present each group with a problem to work out. In a laissez-faire style, you step back and "let go." When students ask you a question or need assistance on solving the problem, you direct them to another group for help or some similar technique that forces the students to come up with solutions on their own. If you have a laissez-faire supervisor, make sure you develop the skills required to take the lead and to assert yourself, when appropriate. The laissez-faire supervisor may actually trust your judgement enough to "let go" of more traditional supervisory tasks, which might include mentoring. While some may think that having a supervisor with this leadership style is "a piece of cake," it actually is not, because the onus is on you to maintain motivation and the momentum of your career. If you need more external motivation, this is another instance in which you might consider seeking a mentor for support, guidance, and a bit of extra motivation from an external source.

Teacher–Leaders

In addition to the classroom setting, you should also know about and develop the qualities of teacher leadership in the wider school environment. Administrators typically cannot provide all of the leadership necessary to resolve the incredibly complex challenges facing schools today. With unprecedented demands placed on the educational system, teacher–leaders can assist with a few projects under the supervision of an administrator. This, of course, depends on your particular school and the issues with which it is dealing at any point in time. Once you become familiar with these challenges, you should start to think creatively about solutions and engage in support activities. This is the first step toward becoming a teacher–leader. Teacher-leaders are experienced teachers and respected role models who have strong interpersonal and communication skills; they take the initiative as influential change agents and have a high level of emotional intelligence. Facilitating collaboration among teachers and administrators, building and strengthening relationships, and empowering teachers are often the means by which teacher–leaders make an impact.

Generally, there are certain critical areas in any school where teacher–leaders can make an impact: as mentors or coaches for teachers in the early stages of their career, or for those teachers who may need support; working with administrators to analyze student data; and on curriculum and assessment projects. As a mentor or a coach, the teacher–leader opens up their classroom to other teachers and encourages a shared dialogue, lesson modeling, and a reflective practice that results in an improved teacher and student performance. A teacher–leader might, for instance, help a teacher better understand the new standards that may be the result of education reform, give feedback on instructional delivery, and help identify up-to-date resources for lesson plans. They might also invite the teacher into their own classroom to observe a particular lesson and its delivery. Through the shared commitment to educational excellence, this collaboration between the teacher–leader and teachers ultimately results in greater student achievement and better outcomes for the school in general.

Another benefit for the school with teacher–leaders is that the administration has a direct connection to what is going on in the classrooms since teacher–leaders are still working in the classroom themselves. When administrators no longer deal with the quotidian classroom issues of their early career days as teachers (if they were classroom teachers), they may feel out of touch with the current classroom challenges. This is through no fault of their own, since administrators are often over-burdened with bureaucratic processes and procedures. With the assistance of the teacher–leader, who can be a representative or an advocate for teachers, administrators have the opportunity to address classroom issues in a timely manner. Problems that may have, in the past, gone unnoticed and therefore unresolved can be brought to the attention of the administrator by the teacher–leader along with, or on behalf of, the teacher or other staff member.

When school administrators unleash the untapped potential of teachers, they find the teacher–leaders who, for instance, may have an expertise (or a graduate degree) in curriculum design or the latest assessment methods. If these are areas in which you had specialized training, make sure your supervisor is aware of it. And, in general, stay current of new developments in education theory and best practices through joining professional organizations—where there also might be leadership opportunities—and reading education academic journals.

Trends in Leadership

There appear to be some interesting trends in how people are thinking about and approaching leadership. One trend identifies a slant toward what has been called a more "feminine"[5] style of leadership.[6] It is possible that this shift addresses some of the changes in the nature of work and in the workforce—where collaboration now seems to be valued over command with regard to leadership style. Others point to a leadership trend toward a respect for "soft skills" such as emotional intelligence and the need to prioritize the "whole" person when it comes to work-life balance, options for remote work, and flexible schedules.[7] This may indicate a preference for coaching, service, or democratic types of leadership styles. For motivated teachers, a supervisor who allows for engagement in decision-making processes and the opportunity to take on leadership positions, might exhibit dominant tendencies in these specific leadership approaches (coaching, servant, democratic). Yet, in the classroom setting where teachers must motivate students of various levels of preparedness to

---

[5] Although, as you will read in Chapter 10, ideas about what is feminine and masculine are expanding, and traditionally-defined feminine values such as collaboration, flexibility, and cooperation are held by men as well.

[6] J. Gerzema & M. D'Antonio, *The Athena Doctrine: How Women (and the Men Who Think Like Them) Will Rule the Future.* San Francisco, CA: Jossey-Bass, 2013.

[7] R. Power, 4 New Trends in Leadership to Watch in 2020. *INC Magazine*, January 28, 2020.

learn, a combination of the various types of leadership styles is the best approach. As you start to consider your potential to be a leader in the field of education, understand your motivational drivers, emotional intelligence level, and communication methods to establish your own personal leadership style. Do not leave this to chance—or to the "default" leadership mode. Once you develop your leadership style, apply or adapt it to meet specific challenges. And, finally, take charge of your professional development as you navigate the transition from a teacher to a teacher–leader, if that is your goal.

Chapter 10

Understanding the Changing Landscape of Gender and Race

Over the course of the last 100 years, there have been a number of critical advances in matters of gender and race. From outdated notions of separate spheres for men and women to the Me-Too movement, and from segregation in the United States to the Black-Lives-Matter movement, we have seen sweeping changes. Yet, as these movements have indicated, there is still much work to be done to ensure that all people have access to quality education, health care, and legal services, among other institutions—no matter their gender or race or sexual orientation. Men and women also need the freedom to express themselves and deserve to be treated with respect no matter their sexual orientation, gender identity, or race. As part of the institution of education, teachers play an extremely important role in creating a classroom environment that is modeled on the concepts and practice of fairness, equity, empathy, and respect. In order to create the best classroom environment for your students, it is your responsibility to be conversant with the latest developments regarding gender and race. In this chapter, definitions of gender identity and sexual orientation terms from the Human Rights Campaign's[1]

---

[1] The Human Rights Campaign is the largest civil rights organization working to achieve equality for lesbian, gay, bi-sexual, transgender, and queer Americans. www.HRC.org.

"Glossary of Terms," will help you understand the new language being used to discuss these topics. In addition, you must also be familiar with the laws as well as individual school policies as they pertain to your students' rights and your rights.

While we explore issues regarding gender and race in this chapter, we also realize the importance of understanding the dynamic between race, gender, sexuality, and social class. To this end, we acknowledge that any discussion of race and gender is better informed by considering how all of these categories intersect. Kimberlé Crenshaw, a Columbia University law professor, introduced the concept of "intersectionality" in 1989. Intersectionality refers to the complex ways that multiple forms of discrimination intersect (or overlap), especially for marginalized groups or individuals (Crenshaw's research focus was on Black women).[2] Over time, the concept of intersectionality went from the field of legal studies to became more mainstream, and it was added to the *Oxford English Dictionary* (OED) in 2015. The OED defines intersectionality as "(t)he interconnected nature of social categorizations such as race, class, and gender, regarded as creating overlapping and interdependent systems of discrimination or disadvantage; a theoretical approach based on such a premise."[3] The material covered in this chapter is situated in an understanding of intersectionality. In other words, to meet the needs of diverse groups of students, we need to address gender equity and racial equity simultaneously (along with sexual orientation and social class, when relevant) and not as discrete or opposing issues. As part of a foundation for this work, we identify the pitfalls in some common ways of thinking and reasoning and provide examples that will help guide you in understanding and discussing these complex issues.

First, it is important to understand some basic concepts that will help avoid misunderstandings or conflicts around issues of gender

---

[2] K. Crenshaw, Demarginalizing the Intersection of Race and Sex: A Black Feminist Critique of Antidiscrimination Doctrine. *Feminist Theory and Antiracist Politics.* University of Chicago Legal Forum, 1989.

[3] *Oxford English Dictionary*, "New Words List, June, 2015." Retrieved on June 9, 2020 at: oed.com.

and race in your classroom, in the school environment, and in general. Misunderstandings or conflicts around issues of gender equity and race equity usually arise when one of these three ways of thinking is in operation: either/or thinking, hierarchical thinking, and universal thinking. The most effective way to avoid these rigid and limited constructs is to adjust to a "both/and" way of thinking about gender and race.[4] It is also important to become familiar with and understand some of the new language that has emerged around gender and race—especially as it relates to identity. You may have heard of the term "non-binary"[5] in your community or through media, for example. Perhaps you have heard someone identify themself as non-binary and they have requested that you use a specific pronoun when addressing them. We all rely on language to communicate our deepest selves, so it is important to honor the language that someone choses to express who they are. Still, you might be confused about the meaning of this term, how to use it correctly, or how to address a person who has identified themself as being non-binary. You are not alone in your confusion. This chapter will examine these two areas—ways of thinking and the use of language—in more detail so that you can feel confident in your ability to handle difficult conversations and situations in your classroom. With a firm grasp of these subjects, you will be able to support those students who may struggle with their identity as well as those who resist inclusive language or thinking. As a teacher, you have tremendous influence on the students in your class, and have the opportunity to support their development into respectful, empathetic, and open-minded citizens.

Either/Or Thinking

You are either a woman or a man, male or female. You are either black or white. You are either heterosexual or homosexual. You must be

---

[4] D.S. Pollard, Perspectives on Race and Gender. *Educational Leadership*, 53(8), 72–74 (1996).

[5] An adjective describing a person who identifies as neither male nor female, rejecting the binary construct historically associated with the definition of gender.

either one or the other. This either/or thinking (or reasoning) when it comes to gender and racial equity sets up a false dichotomy, which is the core of the problem. Either/Or suggests that there are only two options, and you must choose one, not both and not something else outside of the two options. Once you choose one, the other option is intentionally or unintentionally discarded. Depending on the actual situation, being the option not chosen can have consequences. With the either/or dichotomy, there is the assumption of only two possibilities when, in fact, there may be a continuum of possibilities. For instance, either/or thinking might characterize groups in the following way: women or minority? You might see the problem with providing an "either/or" with the categories of women and minority since these categories overlap. Perhaps this is better illustrated through a more elaborate example relevant to gender and race equity in a classroom setting.

Supposing a teacher is planning lessons for the months of February and March. In the United States, February is Black History Month (or African–American History Month) and March is Women's History Month. How would this teacher avoid either/or thinking when creating a lesson plan that focused on the celebration of Black people and women? Which month, for instance, would the American Black abolitionists Harriet Tubman and Sojourner Truth fit in? The teacher might avoid the subject of a special month altogether, but then would lose the opportunity take advantage of the events connected to the month-long celebrations and the chance to discuss major historical figures. Instead the teacher might consider how to breakdown the either/or situation by including women in the celebration for both months. Either/or becomes "both." When talking about the inclusion of both women in both months, the teacher might introduce the concept of false dichotomy to the class, if appropriate. In addition, the teacher might also want to ask the class who is celebrated during the other months of the year? And why do we need a special designated month to focus on women and African Americans? September 15th to October 15th is Hispanic Heritage Month in the United States, and if your school year goes into the month of June, which is Gay & Lesbian Pride Month, you can include those topics in the discussion as well.

A thoughtful teacher might want to mention that although for many years women mathematicians have been ignored to a large extent when looking at the history of mathematics, there are, in fact, important female mathematicians who have made significant contributions to the field of mathematics. For example, the Italian mathematician Maria Gaetana Agnesi (1718–1799), left a legacy with her development of the cubic curve known today as the "Witch of Agnesi," and can be drawn from the equation $y = \frac{a^3}{x^2 + a^2}$. Another female mathematician of note, is the French mathematician Sophie Germain (1776–1831), who made major contributions to the field of number theory. There are quite a few women mathematicians who have made major contributions in the field such as Ada Lovelace (1815–1852), who some claim was the first computer programmer in history. Or the Russian mathematician Sofia Kovalevskaya (1850–1891) who made major contributions to the theory of differential equations, and whose name remains well-known through the Cauchy–Kovalevskaya theorem. The field of abstract algebra highlights the German mathematician Emmy Noether (1882–1935). This should give you some idea of some of the heretofore ignored mathematicians who had to fight during their professional lives for the appropriate recognition that they deserved.

Hierarchical Thinking

Another type of thinking to avoid is hierarchical thinking. In most societies, gender, race, and social class tend to be understood as part of a hierarchy. Children learn at a young age through their family, media, and even religious institutions, where each of these groups fall on that hierarchical ladder. For example, some religions make it clear that they consider homosexuality and bisexuality a sin. Based on the beliefs and values of their parents, children start to categorize various groups according to whether they are influential, powerful, acceptable, or unacceptable. Children notice when one group is privileged over another and when one group ranks higher or lower by virtue of their gender, race, social status, or sexual identity (or a combination of these categories). The beliefs and values that support this hierarchy are also evident in the legal system, in the business world, in medicine, and just about every aspect of society.

Keeping with our prior celebratory-months example, it is clear that the impetus behind the introduction of these months is to give special attention to a category of people who had been historically ignored. By excluding women and people of color from curricula, whether intentionally or not, teachers send the message to students that these groups are not important, they are peripheral to this material, and have not made significant contributions to this field. The effects of hierarchical thinking are multi-generational and can become institutionalized since one generation of scholars pass on what they determine is important and that knowledge or those works become part of the "canon"[6] of that particular discipline. Once teachers understand the deletory effects of hierarchical thinking, they can search for material that is more representative of the diversity of our modern world. This material is critical in creating up-to-date curricula and lesson plans that include contributions from diverse groups of people.

One good example of a diverse mathematician entering the field in a rather unusual fashion is the Indian mathematician Srinivasa Ramanujan (1887–1920) who came from a very poor Indian family and was able to impress the top mathematicians at the University of Cambridge in England with his brilliance. As a result, Ramanujan made significant contributions in mathematics and is still revered by various countries issuing postage stamps with his picture on it and a well-known full-length film *The Man Who Knew Infinity* (Warner Bros., 2016).

Universal Thinking

Disregarding diversity within a group limits our perspective and understanding of that group. Universal thinking results in statements

---

[6] In the field of literature, the canon consists of the classics (Shakespeare, for example, in the Western literary canon). During the past 40 years, certain scholars challenged the canon, arguing that it is a limited survey of literature that left out works from people of color, women, the working classes, and LGBTQ perspectives. These scholars do not want to destroy the canon because those works are important; their goal is to make the canon more inclusive.

that oversimplify broad groups, for example: "All women are ..." or "All Blacks believe ... ." These statements stem from ignorance or from intellectual laziness since we know that not all women or all Blacks are the same. But when we see gender and race categories as universal, we ignore the many differences that exist within that category. This is closely related to stereotypical thinking in that a stereotype improperly relies on the generalization of a particular group of people or, at the very least, a characterization of a group regarding personality, preferences, or ability. A stereotype reduces and simplifies a broad category to just one "type." Universal thinking often leads to sexist and racist assumptions that have contributed to the lack of minority groups in certain fields, for example, since at one point in time it was thought that women and Blacks did not have the intellectual capacity to become engineers, surgeons, business leaders, or pilots—just to name a few White male-dominated fields. Historically, in the United States, women and Blacks have been prevented from enrolling in certain colleges, universities, and programs of study, greatly reducing their opportunities.[7] In the late nineteenth-century and into the early 20th-century, a few prominent physicians in the United States made the argument that the academic rigors of higher education would damage the delicate reproductive system of women.[8] Some colleges and universities instituted quota systems or other restrictions that put a cap on the number of students who did not fit the racial and ethnic profile of their institution.[9] The establishment of historically Black colleges such as Howard (est. 1867); Morehouse (est. 1867), Spelman (est. 1881), and women's colleges such as Smith (est. 1875), Wellesley (est. 1875), and Mount Holyoke (est. 1893), to name just a few, was meant to provide quality educational opportunities to those who were not permitted to enroll at other institutions.

---

[7] *Harvard Business School*, for example, did not admit women until 1963. There are many other similar examples, unfortunately.

[8] E.H. Clark, *Sex in Education; or, A Fair Chance for the Girls*. Boston, MA: J.R. Osgood, 1873. M.A. Lowe, *Looking Good: College Women and Body Image, 1875–1930*. Baltimore, MD: Johns Hopkins University Press, 2003.

[9] J. Karabel, *The Chosen: The Hidden History of Admission and Exclusion at Harvard, Yale, and Princeton*. Boston, MA: Houghton Mifflin Harcourt, 2006.

These exclusionary policies illustrate how universal thinking has historically generated and supported sexist and racist laws and policies.

Despite these daunting challenges, a pioneering African–American woman, Katherine Johnson (1918–2020), became a research mathematician at National Aeronautics and Space Administration (NASA) in the 1950s. According to NASA: "First, she helped put an astronaut into orbit around the earth, and then she helped put a man on the moon . . . [Johnson's] calculations of orbital mechanics were critical to the success of the first and subsequent U.S.-crewed space flights."[10] Johnson was awarded The Presidential Medal of Freedom in 2015, by President Obama, and was also the recipient of a Congressional Gold Medal and the NASA Group Achievement Award. The work of Johnson and her mathematician colleagues at NASA, Mary Jackson (1921–2005) and Dorothy Vaughan (1910–2008), who were also African-American, was featured in the Oscar-nominated film *Hidden Figures* (Fox Pictures, 2016). Before this film was released, few people knew just how critical the work of these three mathematicians was to NASA and to the success of the United States space mission.[11] They were indeed "hidden figures" until just a few years ago.

Both/And Thinking

As an alternative to these three limited ways of thinking and reasoning, the educator Diane Pollard suggests an alternative: Both/And thinking. This type of thinking encourages "the recognition that students, as well as teachers, occupy multiple statuses at the same time

---

[10] Retrieved on July 17, 2020, at: https://www.nasa.gov/content/katherine-johnson-biography.

[11] For more information about Katherine Johnson, Dorothy Vaughan, and Mary Jackson, go to the NASA website where there are wonderful STEM educational resources appropriate for grades K-12. Go to: https://www.nasa.gov/modernfigures/education-resources.

and that these must be taken into account in the school setting."[12] Both/And thinking can be compared to the concept of intersectionality mentioned earlier in the chapter, since intersectionality acknowledges multiple statuses. Pollard also points out that in order to meet the needs of students, we must address the needs of race equity and gender equity simultaneously. Too often the needs or goals of marginalized groups are presented as diametrically opposed, as in the idea that they are two opposing and conflicting points. Addressing issues of gender, race, social class, and sexuality through an intersectional lens or through a Both/And approach (which means acknowledging multiple statuses and the interconnected nature of these social categories), can be the route to a genuine acceptance and celebration of the diversity among the students in your classroom. This is a wonderful opportunity to model ethical practices and creates trust between you and your students in a safe, supportive, and inclusive classroom environment.

New Language for a New World

Gender and sexuality

There are still many stigmas associated with LGBTQ individuals; these groups include lesbian, gay, bi-sexual, transgender, and queer people. Sometimes this acronym also includes intersex and asexual people and will appear as LGBTQIA or LBGTQ+. In a relatively short period of time, the change in this acronym, to make it more inclusive, illustrates how these key terms and definitions are in flux and contested, so keep in mind that these terms and definitions continue to evolve. This section will help you to navigate the new language of gender, race, sexuality, and ethnicity, and will provide current definitions of key terms from the Human Rights Campaign, which could easily expand in the ensuing years. It is important that teachers understand the different labels used within these communities. A lack

---

[12] D.S. Pollard, "Perspectives on Race and Gender. *Educational Leadership*, 53(8), pp. 72–74 (1996).

of understanding, of course, can lead to prejudice and discrimination. In addition, teachers without this knowledge may struggle to understand the full range of identities, which puts them at a disadvantage when it comes to supporting and advocating for their students. Teachers play a critical role in ensuring that all students—including LGBTQ students—are able to learn and explore in a safe classroom space. A basic understanding of the language used to describe various genders and sexual orientations is critical, so we start by defining some key terms.

Gender is no longer understood as "Either/Or": male or female. One of the major changes in how people communicate their gender is illustrated in the variety of relatively new terms that describe this state of being. Gender can now be defined as "both" or "and" or "neither"—in addition to male and female. There are also words to express gender identities. In general, gender is defined as "one's innermost concept of self as male, female, a blend of both or neither—how individuals perceive themselves and what they call themselves. One's gender identity can be the same or different from their sex assigned at birth."[13]

Earlier in this chapter we mentioned the word non-binary. If someone identifies themselves as *non-binary,*[14] they are saying that they are neither male nor female. They reject the binary construct historically associated with the definition of gender. Similarly, someone who says they are *gender fluid* "does not identify with a single fixed gender; of or relating to a person having or expressing a fluid or unfixed gender identity." Gender fluid is sometimes also expressed as *gender expansive.*[15] *Transgender* is an "umbrella term for people

---

[13] All definitions for gender and sexual orientation are from the Human Rights Campaign's (HRC) glossary of terms. www.HRC.org.

[14] Non-binary: "An adjective describing a person who does not identify exclusively as a man or a woman. Non-binary people may identify as being both a man and a woman, somewhere in between, or as falling completely outside these categories. While many also identify as transgender, not all non-binary people do." From HRC's glossary of terms. www.HRC.org.

[15] Gender expansive "conveys a wider, more flexible range of gender identity and/or expression than typically associated with the binary gender system." www.HRC.org.

whose gender identity and/or expression is different from cultural expectations based on the sex they were assigned at birth. Being transgender does not imply any specific sexual orientation. Therefore, transgender people may identify as straight, gay, lesbian, bisexual, etc." *Gender transition* is the process by which some people strive to more closely align their internal knowledge of gender with its outward appearance. Some people socially transition through dressing, using names and pronouns, and/or being socially recognized as another gender. Others undergo physical transitions in which they modify their bodies through medical interventions. Finally, *cisgender* is a term used to describe a person whose gender identity aligns with those typically associated with the sex assigned to them at birth. Cisgender is a word commonly used to replace "heterosexual."[16]

Note how some of the definitions seem almost identical in meaning. For example, *genderqueer* people typically reject notions of "static categories of gender and embrace a fluidity of gender identity and often, though not always, sexual orientation. People who identify as 'genderqueer' may see themselves as being both male and female, neither male nor female or as falling completely outside these categories."[17] Genderqueer may sound much like non-binary and some confusion may arise out of the seemingly indistinguishable terms. However, when speaking with someone who identifies as genderqueer instead of non-binary or gender expansive, we must respect their right to self-identify. One of the most important points regarding this particular subject is that when in doubt as to the meaning of a word or what you should call someone, ask politely for clarification. It is not a disaster if you make a mistake in your use of these terms! Simply correct yourself immediately, apologize, and move on. If one of your students makes a mistake, gently correct them by reminding them of the correct term.

This list of terms and definitions related to the various kinds of genders is not exhaustive. But it will help you to understand why, in

---

[16] Heterosexual and homosexual are outdated terms. It is better to choose gay or lesbian instead of homosexual and cisgender for heterosexual.

[17] www.HRC.org.

addition to naming their lived experience, a person may want to use a different pronoun other than "he" or "she." For instance, a person who identifies as non-binary (neither male nor female), might not want to be referred to as he or she. What pronoun should be used? Most likely, in your English class, you were taught that pronouns are either masculine or feminine, for example, she/her/hers and he/him/his (or plural). Eventually, because of societal changes regarding language use, we tried to avoid these labels because not every "he" felt masculine or male, and not every "she" identified as feminine or female. We have instead turned to using gender-neutral pronouns such as they/them/theirs—which are now acceptable to use in the singular. A few of the new pronouns include: co/co/cos, en/en/ens, yo/yo/yos, ve/vis/ver, and ze (or, xe or zie)/hir/hir, are the most common, although there are a few more.[18] Some people prefer not to use pronouns at all and will use their name in place of a pronoun: "Heather ate a big piece of Heather's cake because Heather has a sweet tooth." An example of one of the new pronouns is: "Heather ate a big piece of *vis* cake because *ve* has a sweet tooth" (using the pronouns ve/vis/ver instead of she/her/hers).

There are also terms that describe sexual orientation (different from gender), and make up the acronym LGBTQ(IA or +). A *lesbian* is a woman who is emotionally, romantically, and sexually attracted to other women. If you are *bisexual*, you are "emotionally, romantically or sexually attracted to more than one sex, gender or gender identity though not necessarily simultaneously, in the same way or to the same degree."[19] *Gay* refers a person who is emotionally, romantically, and sexually attracted to members of the same gender (can be used to refer to a man or woman). *Queer* is a term people often used to express fluid identities and orientations, and is regularly used interchangeably with "LGBTQ." You may occasionally see the addition of an "I" and an "A" to LGBTQIA. The "I" refers to *intersex*, which is an "umbrella term used to describe a wide range of

---

[18] Never refer to a person as "it" or "he-she" since these are offensive slurs against gender non-conforming and trans people.
[19] www.HRC.org.

natural bodily variations. In some cases, these traits are visible at birth, and in others, they are not apparent until puberty. Some chromosomal variations of this type may not be physically apparent at all."[20] The "A" refers to *asexual*, which describes someone who lacks sexual attraction or desire for others. Although it is not a part of the LGBTQIA acronym, you may have heard a person describe themselves as *pansexual*, which is someone who has the potential for emotional, romantic or sexual attraction to people of any gender though not necessarily simultaneously, in the same way or to the same degree.[21]

Race and ethnicity

"What are you?" When posed with this awkward question about identity, we might struggle to find a coherent response. Is the person asking based on physical features or the language you are speaking? If you choose to answer, do you talk about race or ethnicity or even nationality? (some people actually side-step the question by answering "American"). Usually, when someone asks this question, they are really wondering about your race or ethnicity. What exactly is the difference between race and ethnicity? What is the difference, for example, between Latino and Hispanic? While, as an educator, you would never ask a student "what are you?" some of your students might ask this question of you and of other students. Your ability to define race and ethnicity, and to articulate the complex connection between these two words, will help you facilitate conversations about a potentially difficult subject in your classroom.

The United States Census Form (and most government forms) provides the following options in the question that asks about race: white, Black/African American, Asian, American Indian, Alaska Native, Native Hawaiian, and Other Pacific Islander. In the question about ethnicity, there are only two options: Hispanic or Latino, or Not Hispanic or Latino. This is a rather limited way to identify race and

---

[20] *Ibid.*
[21] *Ibid.*

ethnicity, and is problematic for those who do not neatly fit into one of the boxes. So, "What are you?" becomes a much more complicated puzzle.

In general, race might be used to describe (or define) someone by the color of their skin and other physical, social, and biological attributes. However, a more detailed and nuanced definition is needed—one that takes a more historical view of race. Consider this definition of race from the *Encyclopedia Britannica*: "The idea that the human species is divided into distinct groups based on the inherited physical and behavioral differences. Genetic studies in the late twentieth century refuted the existence of biogenetically distinct races, and scholars now argue that 'races' are cultural interventions reflecting specific attitudes and beliefs that were imposed on different populations in the wake of Western European conquests beginning in the fifteenth century."[22] Historically, the concept of race has been used to divide people, often based on superficial physical traits. Consider the example of ethnic groups that are considered part of the white race today: Jewish–, Italian–, Irish–, and Polish–Americans. As newly-landed immigrants to the United States these people were discriminated against. They were viewed as undesirables by the WASP (White, Anglo-Saxon, Protestant) establishment. However, through labor movements, New Deal reforms, and home ownership, these people became part of white America.[23] Unfortunately, this same opportunity for transformation to whiteness (i.e. acceptance) was not possible for Black and African–Americans or other people of color. While it may be a social "construct," in reality, race, determined by skin color, is used to divide people and to discriminate. As we can see by the example of immigrants, race and ethnicity are intimately related.

When we talk about ethnicity, we are often describing language and culture. For example, someone might describe their race as Black and their ethnicity as West Indian, or their race as Native Hawaiian and ethnicity as Polynesian. "Italian" can be both an

---

[22] Retrieved on June 12, 2020, from Britannica.com.

[23] D.R. Roediger, *Working Toward Whiteness: How America's Immigrants Became White, The Strange Journey From Ellis Island to the Suburbs*. New York, NY: Hachette Book Group, 2005.

ethnicity and a nationality.[24] Merriam Webster defines ethnic as "of or relating to large groups of people classed according to common racial, national, tribal, religious, linguistic, or cultural origin or background." In this definition, race is part of the definition of ethnicity.

When talking with your students about race and ethnicity, it is important to have a firm grasp on the inter-related nature of these words as well as a sense of the history of how certain groups of people have been defined. Following the advice from the discussion on gender, always respect the choices of students when it comes to how they define their own race and ethnicity. If you have Latino/Latina students, some may ask that you refer to them as *Latinx*. This is a relatively new word that some have chosen over the traditional gender-identify Latino (masculine) and Latina (feminine).[25] Some of your Black students will prefer to use "Black" as opposed to African American (as is sometimes the case with Caribbean Blacks) and vice versa. Some of your students might not be aware of their race or ethnicity, especially if you teach a predominantly white student population. There are many wonderful educational sources available if you feel you need to delve more deeply into the many complexities of race, gender, sexual orientation, and social class. It bears repeating that when in doubt about how to talk about these issues or how to answer student questions about gender, race, and/or sexual orientation—whether it is about you or them—make sure you understand your school's policy on these matters. Your school may have a counselor available for students who wish to discuss their own development regarding gender and sexual orientation. Hopefully, this chapter has provided a foundation and a better understanding of some basic terms as well as a sense of a new language that reflects the changing nature of race, gender, sexual orientation, and social class today. Remember that it is important to keep abreast of the changes in this field since it continuously remains rather dynamic.

---

[24] This is a good example of "Both/And" thinking mentioned earlier in this chapter.

[25] *Hispanic* refers to people who speak Spanish and/or are descended from Spanish-speaking populations. *Latino/Latina/Latinx* refers to people of Latin American origin or ancestry.

Epilogue

Now that you have completed *Innovative Teaching: Best Practices from Business & Beyond for Mathematics Teachers*, you can incorporate the best practices and techniques we have highlighted throughout when teaching your students, as well as when you enrich their experiences with some of the challenging and fun mathematics examples that encourage student engagement. In addition to incorporating some of the best practices from business and beyond, you can also start to look beyond the world of education for your own supply of best practices that may be relevant to your professional growth.

The main goal of this book is to introduce a particular concept—the concept of innovation in teaching, which, in this case, involves bringing in best practices from fields beyond education to enrich classroom learning for your students. In this way, you borrow and adapt material, techniques, and approaches from disparate fields for use in your classroom. Once you understand this concept, you can start to engage professionals who work in areas outside of education and incorporate new and exciting material. We hope that you use this concept and enjoy discovering what are considered best practices in those areas beyond education as it suits your students' needs and the requirements of your particular school. Teachers have a wonderful opportunity to draw inspiration and resources from practically any source. Those teachers with the curiosity and creativity to identify potential best practices from a variety of sources will inspire their own students to enjoy a lifetime of learning.

Index

A

accountability, 63
active learning technique, 17
administrative processes, 117–118
Agnesi, M. G., 183
agreeableness, 4
algebraic knowledge, 91
Alvesson, M., 61
American public-school system, 168
antibiotics, 107
Arnold, A., 119
asexual, 191
audience
  cognitive issues, 22
  factors, 21
  information collection
    methodologies, 39–41
  relevancy, 21
  students' prior knowledge, 22–39
augmented reality (AR), 138
authoritarian/autocratic leadership, 170

awareness, 2
Aycan, Z., 151

B

Babbage, C., 27
back-stabbing, 162
Bain, R., 137
Bancroft, M., 168
baseball batting averages, 161
Bayes' rule, 82
Bayes, T., 82
Beck, R. C., 60
Bertrand, J., 80
Bezos, J., 167
birthday problem, 36, 122
bisexual, 190
Bitner, J., 143
Bitner, N., 143
Black-Lives-Matter movement, 179
both/and thinking, 186–187
B-17 plane, 104, 118
Bridgman, T., 61
Bridwell-Mitchell, E., 151
budgetary issues, 143

business management, 61, 168
business marketing strategy, 119

C
Cain, S., 13
Cauchy-Kovalevskaya theorem, 183
centroid, 160
Chenault, K., 167
circle of involvement, 51–52, 58
cisgender, 189
civility, 128, 130
Clark, E. H., 185
classroom environment, 13, 45, 52,
    56, 59, 61, 64, 179
classroom profile, 27
classroom setting, 163–164
coaching/transformational style,
    171
coercive/transactional leader, 171
cognitive net, 105
communication skills, 44–45, 126
community-building, 128, 130
competence, 60
concurrency/collinearity, 142
conscientiousness, 4, 10
consensus-building, 172
construction process, 105
Cooley, C. H., 3, 18
Costa, P. T., 4
Crenshaw, K., 180
cross-chords theorem, 88
cyber-bullying, 128

D
D'Antonio, M., 176
data-driven approach, 62
Deary, I. J., 17
decision-making skills, 126

democratic leadership approach,
    172
demographics, 26
Denison, D. R., 151
Dewey, J., 168
discrimination, 188
distribution channels, 43–44
Dweck, C. S., 17–18
dynamic geometry software,
    144–145
dynamic software program, 156,
    159–160

E
educational environment
    analytical thinking, 113–114
    checklist, 109
    circuitous method, 111
    components, 108
    conventional thinking patterns,
        115
    end-of-day activities, 112
    majority-solvers group, 114
    mixture problems, 115–116
    pause point, 113
    problem solving experiences,
        110, 114
    simpler analogous problem, 111
Einstein, A., 138
either/or thinking, 181–183
emotional intelligence, 164, 176
emotional management skills, 126
Encyclopedia Britannica, 192
engagement, 103, 108
enhancer, 63
enthusiast leader, 63–64
ethnicity, 191–193
expert leader, 64

extracurricular programs, 156
extraversion, 4, 10
extraversion, intuition, feeling and
  perceiving (ENFP), 5
extraversion, sensing, feeling and
  judging (ESFJ), 5
extroversion, 10, 12
eye contact, 46–51

F
facial expressions, 55, 58
Fibonacci numbers, 48–50
Field, W., 104
Finkelstein, S., 169
five-factor model, 3–4, 9–10, 15
focus groups, 40–41
Folkman, J., 62, 64
formal/informal teaching approach,
  34, 38–39
Fortune Magazine, 168
Frank, G., 17
Freud, S., 17

G
Garfield, J. A., 27, 95–96
Gauss, C. F., 93–94
Gawande, A., 104, 106–107
gay, 190
gee-whiz moment
    division by zero, 83–84
    geometric series, 82–83
    probability, 80–82
gender
    expansive, 188
    fluid, 188
    identity, 179
    sexuality, 187–191
    transition, 189

gender-neutral pronouns, 190
genderqueer, 189
GeoGebra, 156, 159
Geometer's Sketchpad, 156, 159
Germain, S., 183
Gerzema, J., 176
Giesbrecht, J., 139
Goncz, L., 9
Grove, A., 167

H
Haberlin, S., 163
Hall, E. T., 51–52, 58
hand-held calculator, 139
Harding, W. G., 36
Harlem Children's Zone, 168
Harvard Business Review, 169
Heron's formula, 67–69
Hidden Facebook Groups,
  127–128
Hidden Figures, 186
hierarchical thinking, 183–184
high-school-enrollment rate, 141
Human Rights Campaign, 179, 187
Hyper Text Markup Language
  (HTML), 140

I
ice-breaker, 1
immediate-response technologies,
  140
improvisation, 16
incenter, 160
influencers, 119–121, 125–126,
  129–130
information collection
  methodologies, 39–41
instant connectivity, 140

intercultural communication
theory, 51
International Business Machines
(IBM), 140
intersectionality, 180, 187
intersex, 190
intimate space, 51
introversion, 5, 11

J
Jackson, M., 186
Jobs, S., 172
John, O. P., 4
Johnson, K., 186
Johnson, W., 17
Jung's theory, 4

K
Kanungo, R. N., 151
Kaprekar process, 14
Karabel, J., 185
Kinuthai, W., 139
knee-jerk reaction, 133
Kovalevskaya, S., 183
Krause, S., 137–138
Krulik, S., 65, 171

L
laissez-faire approach, 173–174
Lancaster, J., 138
leadership skills
    business management, 61, 168
    classroom setting, 62
    community consciousness, 129
    critical and creative problem-
        solving, 129
    environment and language,
        121

organizational/institutional
    setting, 169
personalized instruction, 169
research, 62
responsibility, 129
school culture, 152
teacher-leaders, 174–176
trends, 176–177
types, 170–174
work relationships, 167
leadership styles
    classroom, 167
    coaching, servant/democratic
        types, 176
    educators, 169
    feminine, 176
    innovative thinking and action,
        168
leadership types
    authoritarian/autocratic
        approach, 170–171
    coaching/transformational
        style, 171
    coercive/transactional leader,
        171
    democratic approach, 172
    laissez-faire approach,
        173–174
    pace-setter and visionary,
        172–173
    service leader, 173
learning receptivity, 13
lecture-based approach, 2
Lee Strasberg's method, 15
Lehmann, I., 157
lesbian, 190
litigators, 46
logical reasoning, 6

looking-glass self theory, 3
Loud, J., 139
Lovelace, A., 27, 183

M
McCrae, R. R., 4
Machiavellian plotting, 162
McGowan, K., 19
McGreal, R., 139
macro level, 106
magic lantern, 138
Mandela, N., 51
Mann, H., 168
mapping, 102
Marriott, A. S., 153
Marriott International, 153
Marriott, J. W., 153
Marshall, S., 139
massive open online courses
    (MOOCs), 139
mathematical curiosities, 96–99
MBTI, 9–10, 12
media literacy, 29
Meisner technique, 15
Me-Too movement, 179
microcosm, 104
micro level, 106
mind-boggling activity, 97
minimum-distance point, 157
Mishra, A. K., 151
mission/vision statement,
    151–152
motivation
    business management, 61
    definition, 60
    educators, 62
    enthusiast leader, 63–64
    expert leader, 64

extrinsic and intrinsic sources,
    60–61
knowledge, 61
personal style, 61
principled leader, 63
supervisor's leadership skills
    and style, 59
motivational techniques
    gee-whiz! amazing result,
        79–84
    guide students to discover
        pattern, 69–73
    mathematical curiosities,
        96–99
    pertinent story, 92–96
    present students with
        challenge, 75–79
    recreational motivation, 89–92
    sequential achievement, 74–75
    students' knowledge gap,
        66–69
    teacher-made/commercially-
        prepared materials, 99–102
    usefulness of topic (purpose),
        84–89
Mõttus, R., 17
multiplier effect, 120–121, 124,
    126–127, 129
Myers-Briggs Type Inventory, 4
    (*see also* MBTI)
Myers, I. B., 5
Myers, P. B., 5

N
Napoleon's theorem, 156
National Aeronautics and Space
    Administration (NASA), 186
National Science Foundation, 140

neuroticism, 4, 9
New England Journal of Education, 95
Newton, I., 138
New York Times, 124
Noether, E., 183
non-binary, 181, 188, 190
non-positive integer exponents, 69–71
nonverbal cues, 46–47, 55–56

O
office politics, 162
one-to-one correspondence, 102
one-to-one function, 102
one-to-one onto function, 102
online educational technologies, 129, 142
online learning, 125
onto function, 102
openness to experience, 4
organizational culture, 150–151, 165
organizational/institutional setting, 169
organizational silos, 163
orthocenter, 160
Oxford English Dictionary (OED), 180
Oz model, 170

P
pace-setter, 172–173
pansexual, 191
Parade magazine, 124
pedagogical theories, 103
personality
  areas, 18–19
  educational psychology, 4

leadership, 167
preferences/ability, 185
teaching strategy and preferred methodology, 2
traits, 1–4, 17
personal space, 51
Pervin, L. A., 4
Peterson, C., 18
physical proximity
  circle of involvement, 51–52, 58
  intimate space, 51
  invisible zone, 53
  measurement, 54
  nonverbal communication cues, 55, 57
  personal space, 51
  public distance, 51
  social distance, 51–52, 58
  trial judges, 56
Plimpton 322, 27
Polk, J. K., 36
Pollard, D. S., 181, 186–187
Posamentier, A. S., 25, 65, 90, 157, 171
Power, R., 176
prejudice, 188
principled leader, 63
problem-solving technique, 8, 24, 114
product, price, promotion, and place (four Ps), 43
professional development, 142, 177
professional integrity, 171
psychological terminology, 63
public distance, 51
Pythagorean theorem, 27–28, 95–96
Pythagorean triples, 28

Q
quadratic formula, 66–67
queer, 190

R
race, 191–193
Ramanujan, S., 184
recreational motivation
    algebra counterintuitive
        peculiarities, 90–92
    understanding percents, 89–90
Retter, J., 12
Richardson, S., 46
Roediger, D. R., 192
role-playing, 17
Rosenhouse, J., 122
Rosenthal effect, 56, 58
Rosenthal, R., 56

S
Sandberg, S., 167
Scantron system of testing, 139
school culture
    brand recognition and
        consistency, 153
    core beliefs and behaviors, 150
    elements, 154–155
    interpersonal relationships,
        162
    leadership, 152
    mission/vision, 151–152
    navigation strategies, 150
    organization, 150–151
    school public relations, 154
    stakeholders, 154
    visitor, 149
school public relations, 154
school's achievements, 155
self-awareness skills, 126

self-discovering
    counting combinations, 71–73
    non-positive integer exponents,
        69–71
self-selected seating strategy, 53
sequential achievement, 74–75
service leader, 173
sexagesimal system, 27
sexuality, 187–191
sexual orientation, 26, 179–180,
    190
Shafer, L., 151, 154
Siemens, G., 139
simpler analogous problem, 132
Sinha, J. B. P., 151
Snyder, T. D., 141
social awareness skills, 126
social class, 180
social distance, 51–52, 58
social intelligence, 63
social media
    business marketing strategy,
        119
    email/cell phones, 142
    environment and language,
        121
    influencers, 119–121, 125–126,
        129–130
    marketing, 121, 124, 126–127
    school's policy, 128
social networking technologies,
    121, 126, 135
specialization, 106
special quadrilaterals, 74–75
speech-giver, 21
Stanislavski's system, 15
story-telling technique
    Pythagorean theorem, 95–96
    sum of arithmetic series, 93–95

students challenge
  concept of π, 75–76
  order of operations, 76–79
student's enthusiasm, 155
student's home environment, 24
students' knowledge gap
  Heron's formula, 67–69
  quadratic formula, 66–67
students' prior knowledge
  birthday problem, 36
  certainty, 37
  collaborative approach, 35
  considering extreme situations, 24
  current events, 28–29, 32
  demographics, 26
  formal/informal classroom environment, 34
  geographical, social and cultural aspects, 23
  group/individual identity, 27
  infographic, 30
  innovative methods, 22
  latitudinal circle, 34
  one-time nemesis, 30
  personal living experiences, 23
  polyhedral angle, 35
  probability, 36–38
  problem-solving skills, 23
  social and cultural norms, 25
  successive discounts, 31–32
  theorem, 25
  unusual procedure, 31
  worst-case scenario, 24
Sullenberger, C. "Sully," 105

T
target audience, 59
target market, 43, 45

teacher involvement, 141
teacher-leaders, 174–176
teacher-made/commercially-prepared materials
  function concept, 101–102
  similar triangles concept, 100–101
teacher's lounge gossip, 162
teachers' willingness, 141
teaching career, 2
teaching technologies
  ball-point pen, 139
  chalkboard, 137–138, 144
  magic lantern, 138
  overarching theme, 141
  overlapping transparencies, 142
team-building environment, 58
The Mathematics Coach Handbook, 171
The Wizard of Oz, 170
thinking
  both/and, 186–187
  either/or, 181–183
  hierarchical, 183–184
  universal, 184–186
Tierney, J., 124
Time magazine, 140
traditional advertising methods, 120
transactional relationship, 124
transgender, 188–189
trial/error technique, 8, 32
Truth, S., 182
Tubman, H., 182

U
United States Army Air Corps, 104
universal thinking, 184–186

usefulness
    concurrency, angle bisectors of
      triangle, 85–88
    product segments, intersecting
      chords of a circle, 88–89

V
Vaughan, D., 186
Visible Facebook Groups, 127–128
visionary, 62, 172–173
visual aids, 47
vos Savant, M., 124

W
Washington, B. T., 168

Webster, M., 193
Welch, J., 167–168
White, Anglo-Saxon, Protestant
  (WASP), 192
White, S., 16
Willmott, H., 61
Witch of Agnesi, 183
Wolfe, J. S., 52
word-of-mouth advertising, 120
World Health Organization, 106

Z
Zenger, J., 62, 64
Ziglar, Z., 59

Printed in the United States
by Baker & Taylor Publisher Services